Thinking Through Organic Chemistry

First Edition

John C. Hershberger

Arkansas State University

cognella®

SAN DIEGO

Bassim Hamadeh, CEO and Publisher
Seidy Cruz, Acquisitions Editor
Anne Jones, Project Editor
Susana Christie, Senior Developmental Editor
Abbey Hastings, Production Editor
Jess Estrella, Graphic Design
Alexa Lucido, Licensing Manager
Natalie Piccotti, Director of Marketing
Kassie Graves, Senior Vice President, Editorial
Jamie Giganti, Director of Academic Publishing

Cover image: Copyright © 2011 iStockphoto LP/RuslanDashinsky.

ISBN: 978-1-79353-325-8

cognella | ACADEMIC PUBLISHING
3970 Sorrento Valley Blvd., Ste. 500, San Diego, CA 92121

ACTIVE LEARNING

This book has interactive activities available to complement your reading.

Your instructor may have customized the selection of activities available for your unique course. Please check with your professor to verify whether your class will access this content through the Cognella Active Learning portal (http://active.cognella.com) or through your home learning management system.

cognella® | ACTIVE LEARNING content

Contents

Introduction

Organic chemistry has a reputation for being a challenging course. This course is challenging for a variety of reasons. However, it is within your reach to learn the principles necessary to be successful. You can facilitate your learning by combining the development of study strategies and tactics with identifying novel, complex topics and then working problems step-by-step. This is not an organic chemistry textbook. This book, and its associated practice materials, are intended to give you a different perspective on learning organic chemistry. It will help you build a framework for thinking through various problems.

The book is divided into three main parts:

1. **Metacognition and Studying**

2. **Misconceptions : Common Stumbling Blocks**

3. **Making the Grade**

In Part I, we discuss the value of your mindset in learning chemistry. We will also briefly discuss identifying and developing a strategy and tactics to achieve that goal. The strategy we will use is metacognition. Then, we will define high-yield and low-yield study tactics to help make the best use of your valuable study time.

Part II of the book will detail common topics that are more complex or that you may have forgotten from general chemistry.

In Part III, we look at and analyze a number of organic chemistry problems. This generates a base of questions that you can refer to when you study.

Finally, remember that organic chemistry is not impossible. Do your best to stay focused. You got this!

How to Use This Book

Many organic textbooks are roughly organized into a few main portions, with many of the later chapters organized around particular functional groups. This book is organized a bit differently in that first two sections deal with metacognition and misconceptions, and the final part is broken down both by type of question and by functional group.

Part I

Metacognition and Studying

Mindset

Let's discuss mindset as it relates to organic chemistry. Organic chemistry can be intimidating for a variety of reasons. It's an important class if you need it to graduate or to prepare you for a pre-professional exam so you can become a great physician, dentist, scientist, and so forth. We should acknowledge there's a lot riding on your grade. Each exam is stressful, and the score that you get will determine your final grade. Let's think through this. Typically, you want to get as many points as possible on each exam. So, you study, take the exam, and you get your grade back.

Scenario 1:

You didn't get the best grade. Usually the next question is, "What happened? I studied so hard!" We can attribute test performance to various factors. Some factors we have no control over. Maybe there were questions that covered material not covered in class or questions seemed more complex than the practice problems that were given. We may think of factors we can influence such as how long we studied, how many practice problems we did, or if we tried to cram the night before. It's easy to become discouraged and think, "I'm lazy," "I'm not smart enough," or organic chemistry isn't for me." These responses are not true, and if we get caught in this mindset it's easy to end up choosing to drop the class or accept a lower grade. For most of us, our natural abilities are not the factor that limits our success. We can learn how to think through these new concepts in a different way.

Scenario 2:

You aced it. To be fair, let's consider the factors at play here, too. Maybe the test was made up of practice problems or the professor decided to give extra credit or not take off points. Those, again, are things you can't control. How about your preparation and how that influenced your grade? If you did well it's likely that you took the time to learn the principles and apply this new and abstract material to problems. Ultimately, the goal here is to maximize the impact of our efforts by addressing the factors that we can influence. We can't change how the test is written or who writes it, but we can prepare for a variety of questions. We can become confident in knowing how to work through the problems and maximize our score. However, we have to define our goal.

Goal, Strategy, and Tactics

For this book, we will define goal, strategy, and tactics as follows:

- **Goal:** This is the overarching theme, or the big picture. For this course, our common goal is to learn enough to get good grades and become comfortable enough with this information so that we can teach it to another person.
- **Strategy:** This is generally how we reach our goal. Here, we discuss metacognition as a primary strategy for success in organic chemistry.
- **Tactics:** This is about particular actions, the actions we take to actually learn how to solve problems. An example is doing practice problems.

Most of us would say that we study to do well on the exam, but we don't identify what methods or tactics we will use or even what strategy we will use to progress toward our goal. Sometimes, we just study randomly or employ tactics without a strategy and hope for the best. First, identify what changes to your current strategy need to be made. It is difficult to think critically about your own approach to studying, but small adjustments can help in this course and others. You may currently be applying previously learned methods to this course and not seeing desired results. This is similar to deciding spur of the moment to climb a mountain wearing normal shoes. What works for every-day walking may not work for mountain climbing.

I recommend adopting metacognition as a key learning strategy. This strategy involves using a predictable and repeatable cycle. We will become more efficient learners by studying smarter not necessarily harder. Then we can assess our own learning with self-tests or practice questions before taking a higher-stakes exam. Your instructor will usually suggest problems from the book or problem sets they have written. There are also other resources available so that you will have plenty of material to self-test with. By using these self-assessments, you can use metacognition as a strategy to achieve your goal of learning organic chemistry.

Metacognition

When learning new subjects like organic chemistry, "thinking about your thinking" is one of the best methods to maximize your understanding and get a better grade. This will be covered in more detail later. **The basic process of using metacognition involves a cycle of studying, self-testing, and reflection.** The idea is to use low-stakes or no-stakes practice questions to get feedback.

There are a number of unique learning challenges in organic chemistry:

- The concepts are abstract.
- Atoms and electrons can't be seen with the naked eye and are difficult to imagine.
- There is an extensive new vocabulary and nomenclature to learn.
- There is an entirely new system of drawing reactions and structures.

Organic chemistry moves fast! You will be learning a large volume of information at a faster pace than most other classes. There is a lot of ground to cover. Most students have little exposure to organic chemistry before taking the course. The course builds on concepts learned in general chemistry, and new principles are introduced and utilized throughout. In order to answer questions multiple concepts may need to be applied to the same problem. Obtaining clear feedback in organic chemistry can be difficult. One question on an exam can require use of a number of different concepts. You can get clearer feedback by utilizing metacognition as a strategy to assist your studying and improve exam preparation. Data shows that students who attended a lecture and learned about metacognition scored a letter-grade higher in the course versus those who did not.[1]

Initially, consider your current study tactics. Chances are some methods you currently use were acquired by imitating others or trial and error. We should differentiate studying from learning. These two words are closely related but not quite the same, and definitions for this discussion will be given here. To learn is to gain knowledge by studying. To study means to read, memorize facts, and so on. We study new information to learn it. Examinations measure how well you learned the material, not necessarily how much you studied the material.

A practical approach to metacognition is to be mindful about which study tactics you're using and assess their effectiveness. Not all portions of organic chemistry will require the same study tactics or the same amount of time and effort. Some topics require memorization, for example, pKa values, functional group names, definitions of acids and bases, etc. Other topics may require applications of multiple concepts, as well as these questions: What is the product of this reaction? What is the best

[1] Cook, E.; Kennedy, E.; McGuire, S.Y. Effect of Teaching Metacognitive Learning Strategies on Performance in General Chemistry Courses *J. Chem. Ed.* **2013**, *90*, 961 - 967.

reagent? Which carbocation is most stable? There are different levels of learning based on complexity. These have been identified and characterized in Bloom's taxonomy.

Bloom's Taxonomy

Benjamin Bloom, an American educational psychologist, published a book with his coworkers in 1956.[2] He categorized and defined a system and common vocabulary for thinking about the goals of learning. Bloom's taxonomy was revised in 2001,[3] and we will refer to that version. It is best visualized with a graphic of a pyramid (see Figure 1).

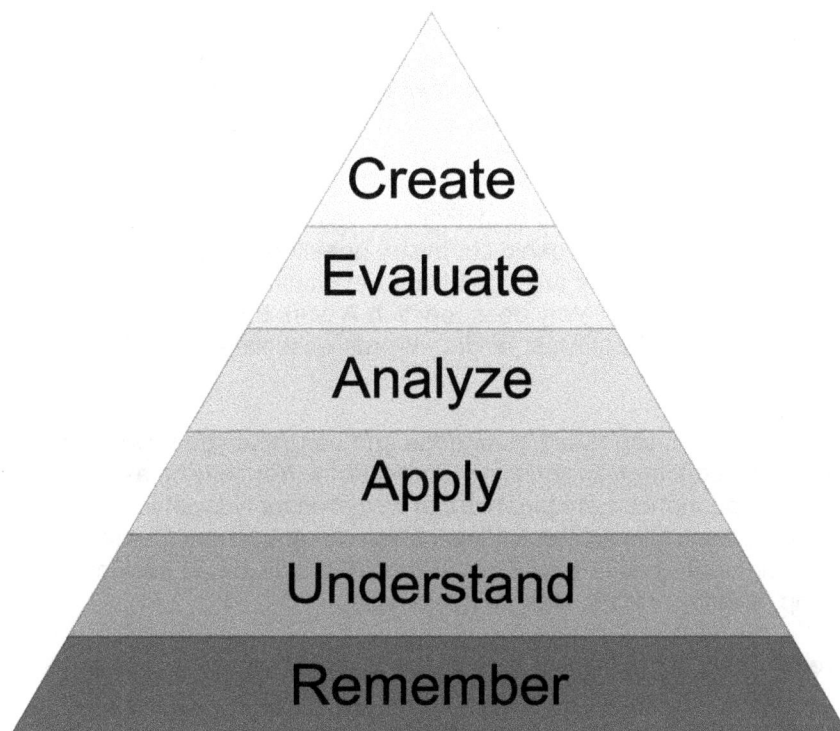

Create

Evaluate

Analyze

Apply

Understand

Remember

Figure 1: Bloom's taxonomy

In the newer version, the levels of the pyramid are labeled with action words to indicate the educational goal achieved. Starting from the base of the pyramid, educational objectives proceed from remembering information to understanding information to applying learned information, and next to analyze the application of information. Levels 3 and 4, applying and analyzing learned information, are generally associated with the undergraduate organic sequence. Levels 5 and 6, to evaluate and create information, are seen in advanced undergraduate and graduate chemistry courses. Sometimes, more challenging questions in undergraduate organic chemistry will have components at levels 5 and 6, requiring evaluation and creation. Let's briefly look at each level in more detail.

---

[2]Bloom, B. S.; Englehart, M. D.; Furst, E. J.; Hill, W. H.; Krathwohl, D. R. *The Taxonomy of Educational Objectives, Handbook I: The Cognitive Domain.* David McKay Co., Inc., 1956.

[3]Anderson, L. W.; Krathwohl, D. R. (Eds.) *A Taxonomy for Learning, Teaching, and Assessing: A Revision of Bloom's Taxonomy of Educational Objectives (complete edition).* Longman, 2001.

- **Level 1: Remember.** Here, we learn how to recall facts and basic concepts or ideas. For organic chemistry, this involves definitions and recalling numbers or equations. Here are some examples. What is a Lewis base? What is ΔG defined as? What is the pKa value of water? Words that you may encounter that would indicate level 1 questions are define, describe, name, order, recall, or reproduce.

- **Level 2: Understand.** Here, we learn how to explain ideas or concepts. In Organic Chemistry I and II, you are asked to identify functional groups, locate the sp^3 hybridized carbons in a molecule, identify which compounds are conjugated, and so forth. Key words that could indicate a level 2 question are explain, summarize, describe, recognize, or give examples of.

- **Level 3: Apply.** The educational objective achieved here involves using concepts to solve a problem. In organic chemistry, this typically involves questions about reactions, for example, predict the product, which reaction will go faster and why, or explain a particular result. A large number of organic chemistry questions involve application of concepts. Words associated with this level include predict, solve, choose, or use.

- **Level 4: Analyze.** This educational objective involves using concepts to compare and contrast information. A large number of organic chemistry questions originate from this level. Examples of questions that require analysis are What is the relationship between two structures? Are they enantiomers or diastereomers? Why does reaction A work but reaction B does not? Order these compounds from most acidic to least acidic. Words associated with this level include analyze, categorize, show, or solve.

- **Level 5: Evaluate.** Here, you utilize knowledge and concepts you have about organic chemistry to make judgments about organic chemistry applications. You may be asked to generate your own synthesis of a compound for example, starting from benzene synthesize benzoic acid. Another example is to use data to give a potential mechanism for a reaction. Questions at this level are not as common during undergraduate courses. Words that indicate an evaluation is required include design, develop, arrange, create, plan, or synthesize.

- **Level 6: Create.** This is the highest level of the taxonomy and involves producing new or original work. Questions that require creation are rare in undergraduate organic chemistry. However, organic chemistry research often utilizes creation because the goal is to generate results from a novel idea. They integrate new data to generate a coherent picture. Questions at this level encourage the utilization of judgment and critique. Sample questions include Which synthesis of Taxol is better and why? How would you improve this synthesis? What would you do differently? Key words include judge, critique, justify, explain, or evaluate.

The Study Cycle

We have discussed the importance of mindset, defined our goal, and explored metacognition as a strategy to optimize performance in this class. Using various tactics we can condense these ideas into a practical application called the study cycle. This practice involves looking and interacting with the material multiple times in a variety of ways. Repetitive recall and use of concepts presented in class via practice problems can help catalyze turning study time into learning gains. We have all tried to cram, and most of us learn that this often leads to poor recall. The study cycle is made up of five steps (see Figure 2.)

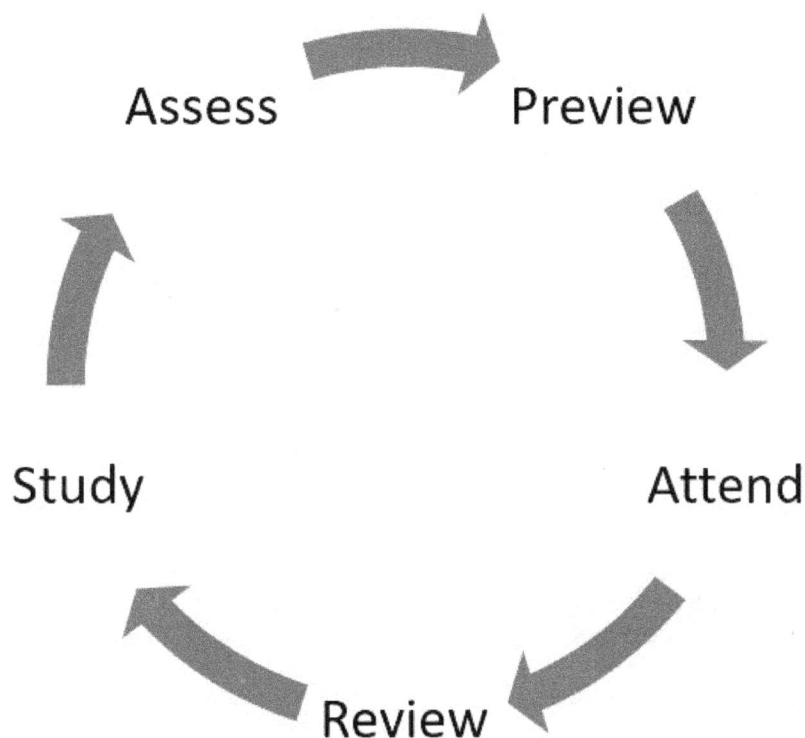

Figure 2: The study cycle

Step 1: Preview. Before class skim the chapter. Note headings and boldface words. Review summaries and chapter objectives, and come up with questions you'd like the lecturer to answer. This should take about 10 minutes. This is the "priming the pump" portion of the cycle.

Step 2: Attend class. Listen mindfully to the lecture, take notes, and ask questions if necessary.

Step 3: Review after class. As soon as you can, reread your notes. Identify any questions. A reasonable amount of time to review is **30 minutes.**

Step 4: Study. Here, repetition is key. Ideally you would have intense study sessions occur three to five times a week and a weekly review in which you read notes and material from that week to make more connections. Here are some tips for studying. Pick a topic for your study session. For example, study alkene reactions or stereochemistry. The topic will guide your effort for the rest of the study session. Try to focus when you interact with the material. You might try organizing your notes, mapping the concepts, summarizing, rereading, filling in your notes, reflecting, and so on. Plan to take regular breaks. Try setting a timer for 15 to 20 minutes and then take a break. End your study session by reviewing everything one more time.

Step 5: Assess. Try the practice problems. This is where you monitor what you are learning. These should be checks on what and how you are studying. The assessments will help you decide if your studying is effective.

Writing and rewriting notes is particularly effective in helping to retain information. **I suggest rewriting your notes at least once before each exam.**

The Learning Equation

For most of us, organic chemistry isn't the only class we are taking, so it is essential that we make the best use of our study time. We can assert the formula

Learning = Time spent x Quality of studying

How much we are going to learn depends not only on how much time we spend studying, but also on how effective our study tactics are. Poor studying is like spending a lot of time doing bicep curls while preparing to run a marathon. While a lot of effort has been expended (poor arms!), it doesn't strengthen the parts of your body that you need to win the race. **Let's explore some common high- and low-yield study tactics.** Table 1 contrasts low-yield tactics with high-yield ones.

Low-yield tactics cause you to lose focus (studying too long, massed practice, blocked practice) or require little effort (reading, rereading, highlighting, and underlining). These tactics are commonly used in high school. At the college level (recall Bloom's taxonomy) their efficacy has diminished. Here are some examples of the high-yield tactics we just discussed.

Comparing and Contrasting Concepts

This tactic can be used to help differentiate two items or ideas. For example, describing the attributes of aldehydes versus ketones, ketones versus esters, *cis* versus *trans* alkenes, nucleophile versus. base, and so on. An easy way to visualize this is with a Venn diagram. Figure 3 below shows an example for aldehydes and ketones.

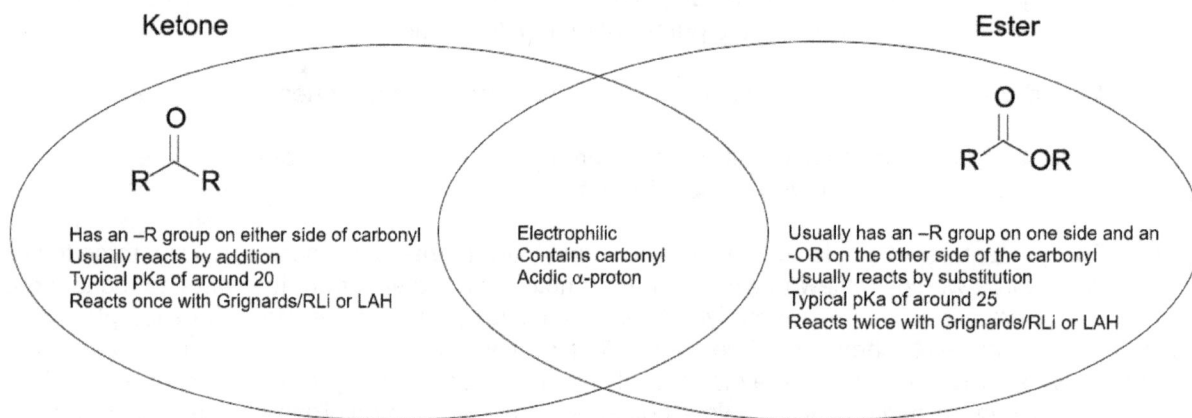

Ketone   Ester

Has an –R group on either side of carbonyl
Usually reacts by addition
Typical pKa of around 20
Reacts once with Grignards/RLi or LAH

Electrophilic
Contains carbonyl
Acidic α-proton

Usually has an –R group on one side and an
-OR on the other side of the carbonyl
Usually reacts by substitution
Typical pKa of around 25
Reacts twice with Grignards/RLi or LAH

Figure 3: Comparison of ketones versus esters

Table 1: High and Low Yield Study Tactics

Study Tactics	
Low Yield	**High Yield**
Studying for too long Your attention span is limited. Most of us can't maintain focus for hours and hours. This is why cramming does not lead to effective learning. It's much more effective to study in shorter increments of time.	**Spacing out practice** Instead of cramming eight hours of study into one night, study one hour a day for eight days. The total amount of studying is the same, but spreading it out allows for more retention.
Massed Practice In this low-yield tactic, you limit study to a single topic and end up repeating the key point over and over.	**Problem variety** Each time you study work on a variety of problems. For example, when studying math, do a mixture of multiplication, division, addition, and subtraction problems versus a block of 50 multiplication, 50 subtraction, and so forth.
Blocked practice This is defined as exclusively reviewing one topic using one method and moving on to another topic using the same method.	**Self-quizzing or testing** This is the highest yield tactic. There are various forms, which include flash cards, practice tests, quizzes, homework problems, and so forth. When you work problems, you are required to recall and then use learned information. This helps with organizing information and concepts that can facilitate the retrieval of information during exams.
Reading, rereading, highlighting, and underlining These tactics are usually not useful on their own. We have all studied ineffectively when using a highlighter and ended up with every line of a page in chartreuse. They are helpful, however, when used in conjunction with other techniques.	**Paraphrase, reflect, and organize information** We can increase effective learning if we manipulate information to formulate connections between concepts and ideas. This includes chapter/concept/reaction maps, comparing/contrasting concepts, Cornell notes, and so on.
	Teach it to others Explain learned information by preparing a mini-lecture or study in a group. This allows you to see and experience the material in a different way.

Concept Map

Figure 4 is an example of mapping out isomers. At the top is the general topic of isomers. Isomers are any compounds that have the same molecular formula but are different from one another. We can divide isomers into two types: constitutional isomers (the way the atoms are bonded to each other is different) and stereoisomers (bonds are the same but have a different spatial orientation). We can further divide stereoisomers into two categories. Enantiomers are mirror images of each other, and diastereomers are all other stereoisomers. Diastereomers are further split between configurational and geometric (e.g., cis/trans) isomers.

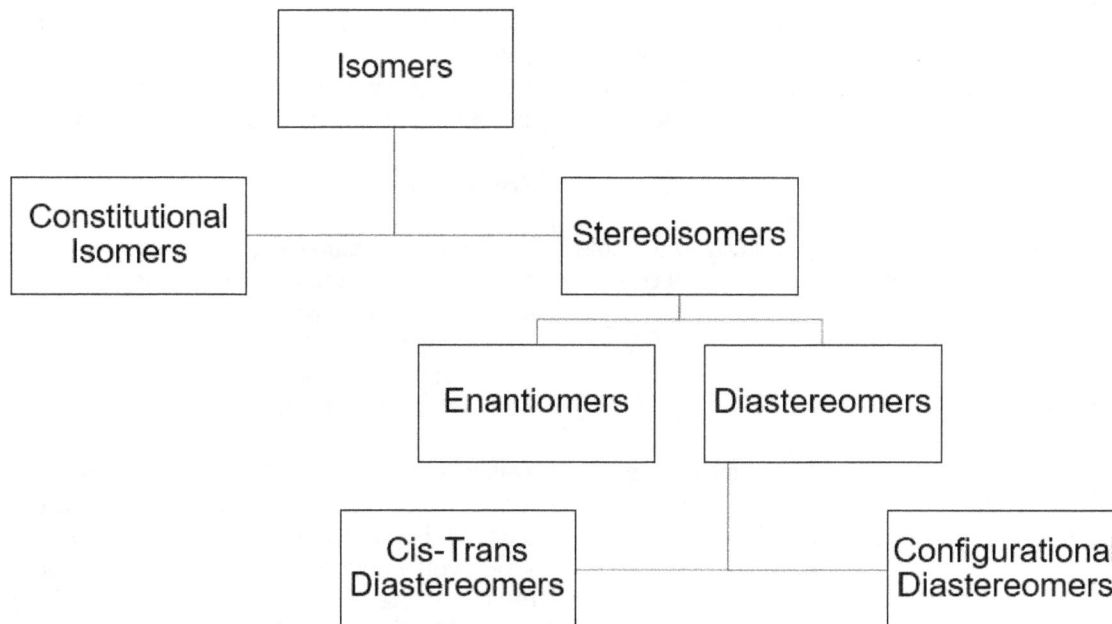

Figure 4: An example concept map for isomers

Cornell Notes

We can use Cornell Notes[4] to make notes more organized and clear. They also help you interact with and review the material more thoroughly. You can either take notes using this format during class or generate Cornell Notes from a set of notes taken in class. An example is shown in Figure 5.

The page is broken up into four main areas:

- **The top part** has subject, chapter, and date.
- **The middle three-quarters of the page** is broken into two sections.
- **The left third of the page** is space for questions that you may generate from the notes.
- **The right two-thirds of the page** is space for the notes taken during class.
- **The bottom quarter of the page** is set aside for a summary of the notes.

Cornell Notes are particularly helpful because they guide note taking and review. They help when rewriting and condensing notes as well. The more you interact with the material the better.

Flash Cards

Flash cards are especially useful when studying relatively straightforward questions like definitions, reactions, concepts, and so on. Making flash cards encourages you to paraphrase, reflect upon, and organize the material. Here are some tips for making flash cards:

[4]Paulk, W.; Owens, R. J. Q. *How to Study in College, 10th* ed; Wadsworth, Boston, 2010.

Organic Chemistry I – Isomers

What are isomers?	Isomers are compounds that have the same composition or molecular formula yet are different in some fashion.
What are the types of isomers?	There are two main types of isomers—constitutional isomers and stereoisomers. Constitutional isomers differ in their bonding, resulting in either differing functional groups (i.e., dimethyl ether versus ethanol—both have the same formula, yet have differing functional groups. Another option is to have a functional group in a different location (propan-1-ol versus propan-2-ol). Stereoisomers differ in the spatial arrangement of the atoms, giving different stereoisomers (R versus S, E versus Z).
How do enantiomers and diastereomers differ?	Stereoisomers can further be divided into enantiomers and diastereomers. Enantiomers are nonsuperimposable mirror images of each other, such as left versus right hands. All other stereoisomers fall into the diastereomer category, and this includes alkene isomers.

Isomers are compounds that have the same formula but differ in their physical or chemical properties. They can be grouped into two different groups, constitutional isomers and stereoisomers. Stereoisomers can be broken into two different buckets– enantiomers and diastereomers.

Figure 5: Cornell Notes

- **Include words and pictures.** The molecular structures are pictures, but feel free to add words or other pictures to help you remember.
- **Add just one question or concept per card.** If you put too many concept on a card, it might make it more difficult to recall everything on the card. If a concept is complex (e.g., isomers or stereoisomers), break it into parts and put each part on a different flashcard.
- **Say the answer out loud before looking at the back of the card.** Make an attempt to recall the information even if you don't know the answer.
- **Study the flash cards in both forward and reverse sequences.**

An example of a front and back for a flash card is shown in Figure 6.

The front shows what the flash card is covering, and the general reaction scheme. Note how esters are generally limited to both R and R' being carbon-based groups. The back of the card has the products, which are a primary alcohol and the -OR' part of the ester as an alcohol.

Homework/Practice Problem Hints

Here are some general hints for problem solving:

- Try to **go through the problems as soon as they are assigned.** This gives you time to space

FRONT

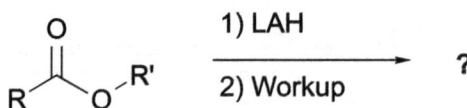

BACK

Figure 6: Example flash card

out your studying, which is a high-yield tactic.
- **If you have access to the answers, do not look at them right away.** Generate an answer even if you know it is wrong.
- **Give yourself at least two minutes of effort on a problem** before you answer.
- **Try to identify the concepts** the problem is asking you to use.
- **If you get stuck, ask others about it.** Chances are that if you have a question, a classmate does too. You may find a tutor at the tutoring center or may choose to ask a classmate or your instructor.

Summary

So far, we have talked about the importance of your mindset when it comes to working your way through organic chemistry. The key here is to remember that you can be mindful and influence your study habits and grade.

Then we talked about our goal, a good strategy, and various tactics. Identifying a goal focuses your effort. Here, the goal is to learn the material so well that you can teach it to others. Metacognition is a great strategy to ensure you become aware of how well you understand the material. The point here is completing assessments to gauge your progress.

We discussed Bloom's taxonomy and its application to organic chemistry. We can see that organic chemistry seems a bit more difficult because the work requires a more complex level of thought. Being able to identify these levels can deepen your learning.

Next, we discussed the study cycle. Preview the lesson, attend class, take notes, and then review after. Try to study three to five times a week. Sticking to the study cycle will increase your chance for success. Figure out what high-yield tactics work best for you while studying and set aside enough time to thoroughly understand the lesson.

The next two sections will include more detail about where you might get confused and then help you think through various organic chemistry problems. Good luck!

Part II

Misconceptions : Common Stumbling Blocks

Preview of Part II

This part of the book deals with a select number of organic chemistry ideas or concepts that are one or more of the following:

1. Not covered in class because of time constraints in many organic courses
2. Assumed students will pick it up along the way
3. Introduces concepts or ideas that are in the background of other courses besides organic chemistry
4. Commonly misunderstood concepts that are important and lead to lower grades if not understood properly.

Drawing and Interpreting Structures

Much organic chemistry information is presented with drawings and structures. This allows for information to be presented in an efficient way but comes at the price of lack of details or other information. Being able to look at and interpret structures of organic molecules is a very useful skill. The drawings of structures also begins very early in chemistry, with some simple organic molecules shown in general chemistry courses. Organic compounds are three-dimensional, and a sheet of paper or a computer screen is only a two-dimensional space.

Thus, there is some information that has to be changed or removed in order to represent a three-dimensional molecule in a two-dimensional space. We will discuss the commonly used skeletal formula. This has the advantage of being relatively easy and quick to draw while giving a number of structural details without overwhelming the eye. The downside is that there are a number of conventions and omissions that must be recalled. Let's look at the structure of a couple of common pharmaceuticals: the non-steroidal anti-inflammatory drug (NSAID), sodium (2S)-2-(6-methoxynaphthalen-2-yl)propanoate, or Naproxen, and an antihistamine, loratidine (see Figure 1).

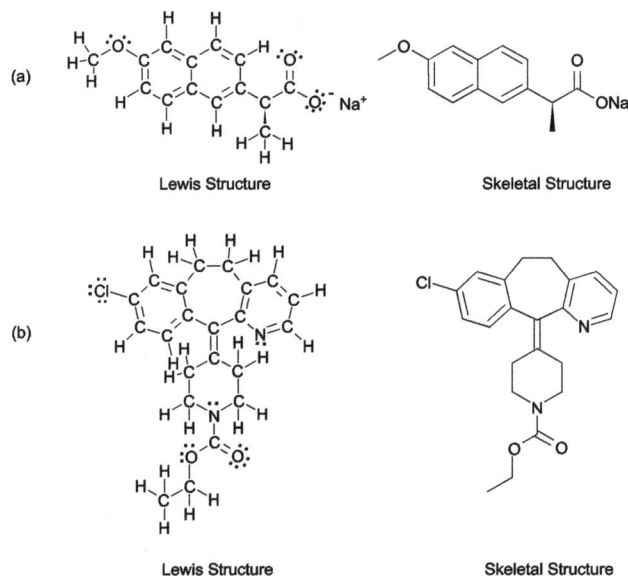

Figure 1: Lewis and skeletal structures

On the left, we have a Lewis structure that has all of the atoms and the lone pairs and charges explicitly drawn on the page. This is visually very busy and can be difficult to look at because of how busy it looks. Each line represents a bond with two electrons. Nothing is really left to the imagination with this structure. On the right, we have the skeletal structure that has the *least* information explicitly drawn in, so we will need to fill in the blanks. Notice how much cleaner the structure on the right looks compared to that on the left. However, they both describe the same compound!

As we can see on the right, each point where a line changes direction or ends is a carbon atom. Atoms that are not carbon or hydrogen (denoted as heteroatoms) are explicitly drawn, and protons on heteroatoms are also explicitly indicated. Except for the C-H on an aldehyde, the protons on carbons are generally not explicitly indicated, and it is assumed that there are enough protons on the carbon to get the number of bonds to four. We also see an example of the dash-wedge convention used at the carbon adjacent to the carboxylic acid. If a substituent is dashed, it is behind/underneath the plane of the surface you are looking at. If is a wedge, it is in front of/above the plane of the surface you are looking at. Being able to visualize the molecules in 3D is incredibly useful.

Acids and Bases

Of the commonly seen organic reactions, acid-base reactions (in the form of protonation-deprotonation) reactions are second only in speed to single electron transfer reactions (such as redox reactions seen in general chemistry) due in part to the principle of least nuclear motion, which states that the less change in atomic position and electronic configuration a reaction causes, the faster it will go.[5] Here we are talking about a *class* of reactions. Unfortunately, each set of conditions gives different competing reactions, and thus each must be evaluated individually. The take-home message is that typically when an acid=base reaction is competing with most other types of reactions, the acid-base reaction will win.

The concept of acids and bases is a foundational one for organic chemistry. A significant number of reactions and concepts in organic chemistry are related to and impacted by acidity and basicity. Many organic reactions involve mechanistic steps that involve deprotonation and protonation. Thus, knowing which protons are most acidic can help predict reactivity. Acids and bases are also related to energy and equilibrium and are a reflection of the structure and environment the compound is in. We will start with a review of the three main concepts of acids and bases, and then discuss how they relate to and are used to understand and predict outcomes relevant to organic chemistry.

The first concept is the Arrhenius concept of acids and bases. These are based heavily on water-soluble compounds. The Arrhenius definition of an acid is a compound that dissolves (typically in water) to give H^+ ions; bases are those that dissolve in water to give OH^- ions. The strong mineral acids (HCl, HNO_3, H_2SO_4, HBr, etc.) are Arrhenius acids, while the alkali metal hydroxides (NaOH, LiOH, KOH) are the most common group of Arrhenius bases. This is the most restrictive concept and is not used widely today because of those restrictions.

The second concept and the one we most often mean when we talk about acidity of organic compounds is the Bronsted-Lowry (or B-L) acid-base concept. Here, the concept is proton-focused; that is, the definition depends on what is happening to a proton. For example, a B-L acid is a *proton* donor, while a B-L base is a proton *acceptor*. This is much more broad since there is no requirement for dissolution, only that there is a proton. Since most organic compounds have a proton, most are B-L acids. What is different is the degree to which they are acids (i.e., are they *strong* or *weak* acids?) B-L bases are proton acceptors, which means that there needs to be some electron density available to accept the

[5]Hine, J. *The Principle of Least Nuclear Motion. Advances in Physical Organic Chemistry*, Vol. 15, Academic Press, 1977, pp. 1 - 61.

proton. The two sources of electron density are lone pairs of electrons and π-bonds. σ-bonds are also very common in organic compounds, but the nature of the σ-bonds being between two nuclei mean that those electrons are typically *not* going to interact with most acids. Not all organic compounds have π-bonds or lone pairs. Thus, virtually all organic compounds can be B-L acids, but there are significant portions of organic compounds (mainly hydrocarbons) that are not practical B-L bases. Examples of compounds that can be B-L bases are shown in Figure 2 with arrows indicating the basic sites.

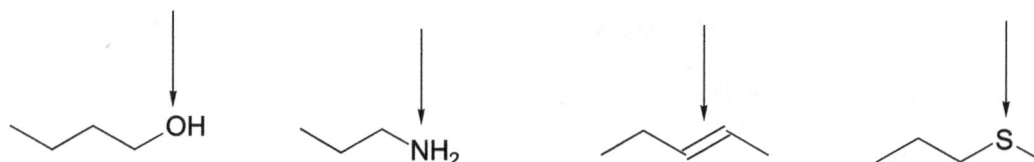

Figure 2: Lewis bases along with indicated basic sites

The third and broadest acid-base concept is the Lewis acid-base concept. This concept revolves around electrons instead of protons. Thus, it is different than B-L or Arrhenius since it has nothing to do with protonss. Lewis acids are *electron* acceptors, while Lewis bases are *electron* donors. This definition now encompasses compounds that did not fit into either previous definition. Lewis acids can be any atom or other entity than can accept electron density. Thus, an acid like HCl is an acid under all three definitions, and a compound like NaOH is a base under all of the definitions. Compounds that contain a group 3 element (boron or aluminum in particular) can be Lewis acids without having any protons in the formula. Lewis bases are quite broad and can be compounds that have a lone pair or π-bond. Figure 3 shows some examples of Lewis acids and bases.

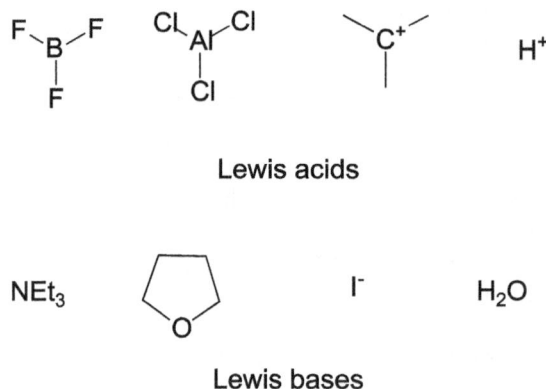

Lewis acids

Lewis bases

Figure 3: Assorted Lewis acids and bases

There are a number of questions that can arise involving acids and bases since the concept is quite broad, and it is a central concept in organic chemistry. These revolve around Bronsted-Lowry acids and bases and Lewis acids and bases. Typical questions around identifying B-L acids and bases involve where the equlibrium lies for a particlar acid-base reaction, rationalizing acidities of organic compounds, ranking similar compounds in terms of acidity, and so forth.

Cycloalkanes

Unlike linear alkanes, cyclic alkanes are less free to move. An illustration of this is to move your arms at all the different angles you can. Notice how you have a lot of angles and freedom of movement. Now clasp your hands together and move them around as you can without opening your hands. Notice that the movement is now more restricted. Cyclic alkanes are similarly restricted but are different based on the ring size. The ring sizes we will look at are three-, four-, five-, and six-membered rings.

As with conformations of alkanes, cycloalkanes can also have torsional and steric strain. However, cycloalkanes also have something called angle strain, which is the strain energy related to the bond angles being geometrically constrained and away from the ideal angle (for us, the 109.5 we see for Csp^3 carbons). We call the sum of the strain energies ring strain. In order to get a more even comparison, the combustion energies are calculated per CH_2 group. Table 1 has these energies for ring sizes 3 to 16 in kcal/mol.[6]

Table 1: Ring Strain per CH_2 Group for 3 to 16-Membered Rings

Ring Size	Strain kcal/mol
3	27.5
4	26.3
5	6.2
6	0.1
7	6.2
8	9.7
9	12.6
10	12.4
11	11.3
12	4.1
13	5.2
14	1.9
15	1.9
16	2.0

As with a lot of real-world examples we see in organic chemistry, there are multiple factors at play that interact to give the experimental ring strain. For the ring sizes three and four, angle strain is a very large factor, since the angles of cyclopropane and cyclobutane are roughly 60 and 90 degrees, respectively. These are both large deviations from the ideal 109.5 bond angle. As the ring size increases, the angle strain and the eclipsing strain both decrease to where they are at a minimum with cyclohexane. As the ring size again increases from 7 to 11, there is an increasing amount of steric strain, similar to the 1,3-diaxial interactions for cyclohexane, but across the ring. Thus, these are called *transannular* interactions. It should be noted that many organic textbooks have a value of zero for cyclohexane; this table is very close at 0.1 kcal/mol.

Cyclopropanes

The three-membered ring of cyclopropane is the smallest possible ring size. The structure of cyclopropanes dictate that all the bonds must be eclipsed as well as have a 60-degree bond angle. Thus, it is no surprise that these are the most strained of all cycloalkanes. In fact, cyclopropanes can be hydrogenated with hydrogen and a metal catalyst! In addition, the oxygen analogs of cyclopropanes (epoxides, or oxiranes) also show an increased reactivity because of this ring strain. Examples are shown in Figure 4.

---

[6]Anslyn, E. V.; Dougherty, D.A. *Modern Physical Organic Chemistry*, University Science, 2006.

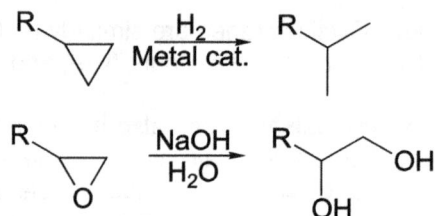

Figure 4: Examples of reactivity because of ring strain in three-membered rings

Cyclobutanes and Cyclopentanes

Cyclobutanes are also very strained, but with a four-membered ring a cyclobutane can contort and twist a bit to make the molecule a bit lower in energy. Cyclopentanes are freer still, resulting in a lower energy system that looks like an envelope, with four carbons roughly in a plane and one out of the plane. Pictures of these can be found in Figure 5.

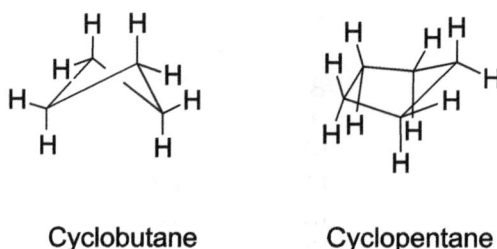

Cyclobutane   Cyclopentane

Figure 5: Cyclobutane and cyclopentane conformations

Cyclohexanes

Cyclohexanes are virutally strainless because the cyclohexane ring can exist in one of two chair conformations, much like a lawn chair. For cyclohexane itself, there are 12 protons, and due to the way the chair conformations exist, there are two types of protons– axial and equatorial. One key point to make here is that there is no relationship between a substituent that is drawn as a dash or a wedge and whether that substituent is axial or equatorial. This is because there are two chair conformations and each substituent will get a chance to be axial in one chair, and equatorial in the other. This is a common assumption that often leads to issues with drawing chair conformations correctly. For cyclohexane, the two chairs with the substituents labeled are shown in Figure 6.

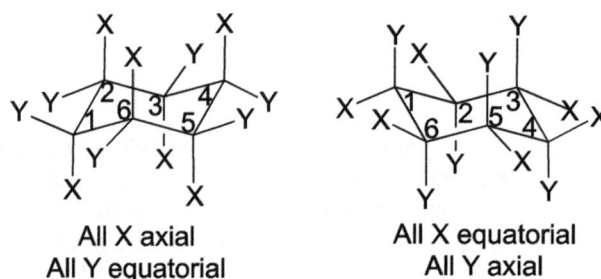

All X axial   All X equatorial
All Y equatorial   All Y axial

Figure 6: Ring conformations

Note how the substituents labeled X are axial and those labeled Y are equatorial in the first chair. In the second chair, those are reversed, with X being equatorial and Y being axial. Note that the axial substituents are drawn vertically up or down, and the equatorial are at a slant, also up or down. A final feature that is important to realize is that the substituent that will be a wedge is drawn higher up on the page than the one that would be a dash. For instance, on carbon 1 the Y is drawn higher up on the page in *both* chairs. As a real example, we will use trans-1,4-dimethylcyclohexane. There are many ways to deal with cyclohexane chairs, and I share the way I draw them in Figure 7.

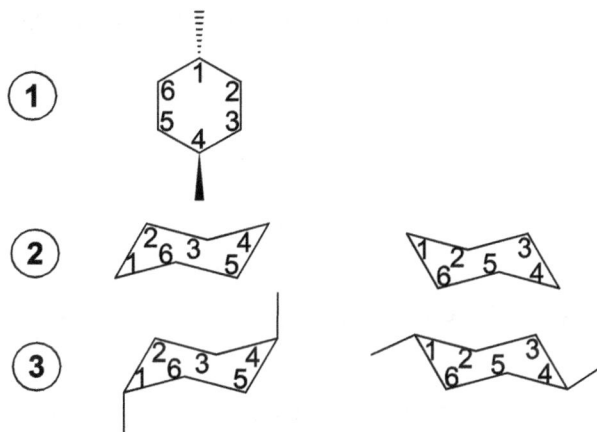

Figure 7: Method to draw cyclohexane chair conformations

1. **Label the carbons.** First, I label the carbons 1 – 6. Whether you label clockwise or counterclockwise does not matter, but that direction needs to be continued when you draw your chair conformations.

2. **Draw the chair conformations and label the carbons.**   There are two chair conformations that can be drawn-one with the left-most carbon pointing down and one with the left-most carbon pointing up. For the chair conformation on the left, the carbons numbered 2, 4, and 6 are up, drawn above the carbons numbered 1,3, and 5, which we will denote as down. Axial is up on the up carbons and down on the down carbons. The chair on the right is the opposite, with the carbons numbered 1,3, and 5 drawn higher than the carbons numbered 2,4, and 6. I typically start with the chair that has the left-most carbon pointing down as carbon 1. That carbon will have the axial substituent pointing straight down.

3. **Draw the subtituents in.** Begin with carbon 1 and place substituents as you work around the ring. The position of the carbon (relatively lower or higher) will guide where the axial and substituents go for that particular carbon. Another challenging aspect of drawing cyclohexane chair conformations is that they can look quite different depending on where each atom is drawn. This is another reason labeling is key. Drawing in both substituents on carbons that are substituted can help make sure the bond angles are correct. Equatorial substituents can be a little more challenging to draw. To help with that, imagine there is a dividing line in the chairs (see Figure 8).

Equatorial groups on the *left* side of the line point left, and those on the right side point right. It should be noted that they point toward the right whether they are going up or down. If the groups are drawn poorly such that they are not easily observed to be axial or equatorial, they are ambiguous and an incorrect representation of the chair. Figure 9 has an example of a bad drawing; note how the bromine, chlorine, and fluorine are not drawn either axial or equatorial.

Rings that have a single substituent will have one chair conformer that is more stable than the other. The conformer with the axial substituent is less stable than the one with the equatorial due to steric concerns. Values have been calculated by observing equilibria between the two chair conformations.

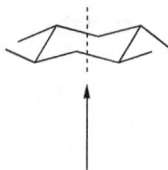

To the left of this line, equatorial substituents point left
To the right of this line, equatorial substituents point right

Figure 8: "Continental" divide for direction of equatorial substituents

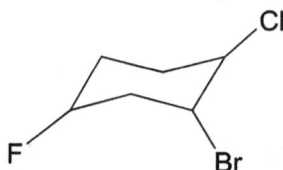

Figure 9: How *not* to draw chair substituents

Table 2 has a selection of data from the Reich compilation of data for the A-Values for the equilibrium shown in Figure 10.[7]

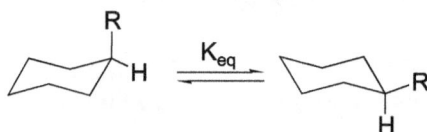

Figure 10: Equilibrium for monosubstituted cyclohexane chair conformations

Table 2: A-Values (-RT log(K_{eq})for Substituents on Monosubstituted Cyclohexane Rings

Substituent	A-Value kcal/mol
Me	1.70
Cl*	0.49
CO_2Me*	1.29
CN*	0.21
OMe*	0.68
Ph	3.0
i-Pr	2.15
t-butyl	>4.5

Note: Starred substituents are averages of more than one value.

Of note is that for a ring that has a *t*-butyl group, since the group is so large (other A-Value compilations have it at greater than 10 kcal/mol) it effectively "locks" the ring into the conformation where the group is equatorial.

---

[7] Reich Collection, *Fundamentals of Organic Chemistry*, https://organicchemistrydata.org/hansreich/resources/fundamentals/?page=a_values/

Stereochemistry and Isomers

Why are isomers important?

1. **Isomers will often have different biological effects.** One isomer may have a useful biological effect, while another may be poisonous.

2. **Mixtures of isomers are often generated during synthetic reactions.** Understanding how to analyze them can be a useful skill.

3. **Isomers have different physical properties.** Being able to use this to separate isomers can be invaluable. Also, predicting which isomers will be produced may allow for reduced waste generation during purification, leading to better environmental outcomes.

Isomers are any compounds that have the same chemical formula but are different in some fashion. Isomers fall into two large categories: constitutional isomers and stereoisomers. Constitutional isomers have a different connectivity; that is if you drew a map of which atoms are bonded to each other, they would be different for a pair of constitutional isomers. Constitutional isomers will have different melting points and boiling points. We can see some examples of constitutional isomers in Figure 11.

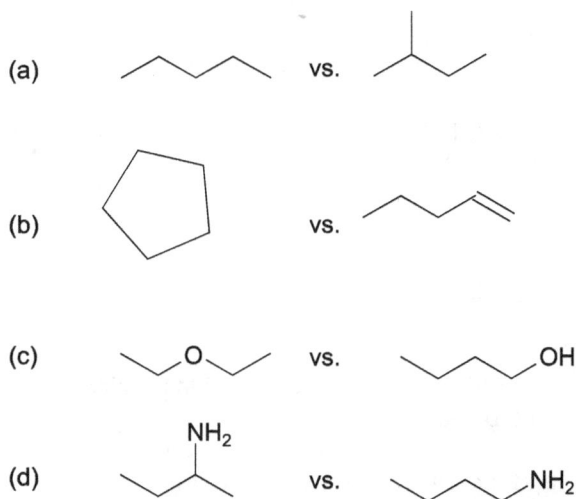

Figure 11: Examples of constitutional isomers

In (a) we have a pair of constitutional isomers where one is linear and one is branched, both with five carbons and the formula C_5H_{12}. In (b) we have a pair of constitutional isomers where one of the isomers is a ring, and the other has a double bond. It is useful to note here that a double bond or an alkene both have one unsaturation; these compounds both have the formula C_5H_{10} instead of the C_5H_{12}. In (c) we have two compounds that have the same formula but two different functional groups: an ether for the compound on the left and an alcohol in the compound on the right. In (d) we have the same functional group, a primary amine, attached to a different areas (or regions) of the molecule. Thus, this pair of isomers are also called regioisomers. Constitutional isomers of all types will have different physical properties, depending on the intermolecular forces metioned in the melting points and boiling points section.

In contrast, stereoisomers will have the same connectivity but different orientations in space. Stereoisomers are further divided into two groups: enantiomers, which are nonsuperimposable mirror images of one another, and diastereomers, which are all stereoisomers that are not enantiomers. A clarification is in order, namely the difference between stereocenter and chiral center. For purposes of this discussion, stereocenters are any carbon at which switching two groups will generate a stereoisomer. A chiral center is an atom (for our purposes, tetrahedral) where switching two groups will change the configuration at that atom. All chiral centers are also stereocenters, but not all stereocenters are chiral centers. For instance, in alkenes that have an E/Z designation, switching groups on one of the carbons will generate the other stereoisomer. However, alkene carbons, being sp^2, are not chiral centers. Examples of these are shown in Figure 12.

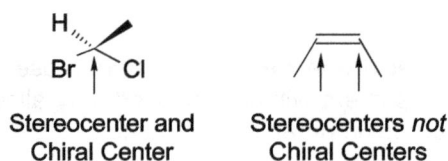

Stereocenter and
Chiral Center

Stereocenters *not*
Chiral Centers

Figure 12: Stereocenter that is also a chiral center versus stereocenter that is *not* a chiral center.

Another concept that is useful to review is that of chirality. Chirality is a property of an object, in our case, an entire molecule. If an object is chiral, then it has a nonsuperimposable mirror image; the classic examples are our hands. The left and right hands are non superimposable to one another. The distinction between a chiral center and an object being chiral is important due to the presence of *meso* compounds, which contain chiral centers but are not chiral objects. Other every-day objects that are chiral include screws, drill bits, and gloves. Carbons are a generally only a chiral center if they are sp^3 hybridized and have four different substituents. Now that we have established that carbons can be chiral, we need a way to describe which 'hand' a carbon is. IUPAC (International Union of Pure and Applied Chemistry) has a method that involves the following steps:

1. Assign priority to the four different groups. This is done by starting from the carbon and looking at the atomic number of each atom that is directly attached. The highest atomic number gets assigned priority 1, the next is number 2, and so on. If a carbon has a hydrogen, it is usually number 4 since it has atomic number 1.

2. The IUPAC method indicates that if the atom/group that is priority four is not in the back, to reorient the molecule such that 4 is in the back.

3. If the groups labeled 1,2, and 3 form a clockwise circle, then the center is assigned the designation *R*; if it is counterclockwise, it is assigned *S*.

This sounds simple, but there are a few issues that come up with these assignments:

A) Assigning priorities is a little more complex if there is a tie in the atomic number of directly attached atoms. If there is an initial tie, there is a tie-breaking procedure that occurs by continuing to go to the next set of atoms and look for a point of difference. There is a fairly common misconception that molecular weight of the entire branch is what determines the priority of that group; this is not the case. A few examples can be seen in Figure 13.

In (a) all four are different, so we simply go by atomic number. The O of the -OH is 1, the F is 2, the carbon of the methyl group is 3, and the H is 4. In (b), we have a good example of working from the center and working outward. Asssigning 1 and 2 is easy, since they both have a higher atomic number versus carbon, so O is 1 and N is 2. Note that it is the atom that is directly attached to the chiral center that is looked at first. At the first level, there is not a difference, since it is C versus C. Then, we go to the next layer of atoms. The carbon that is labeled 3 has the following three substituents: C, C, and

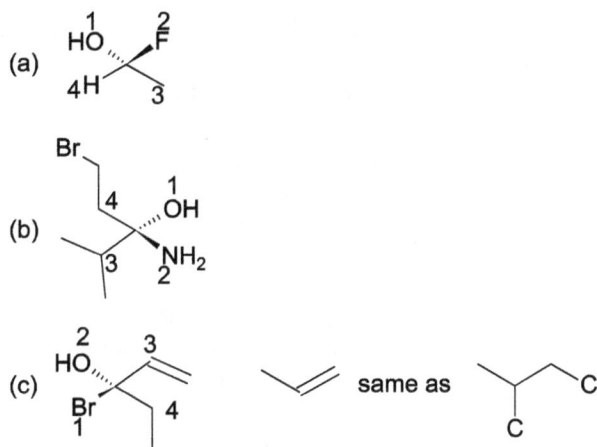

Figure 13: Examples of assigning priorities for designation of *R* and *S* configuration

H. The carbon that ends up being labeled 4 has C, H, and H as the substituents. Thus, the first point of difference is that the carbon labeled 3 has two carbons attached to it and the carbon labeled 4 does not. Also note that the bromine atom does not enter the discussion here. In (c) we now have a multiple bond to deal with. This can be problematic since while there are four bonds to the atom labeled 3, there are only two other atoms attached. We deal with this by at the first layer, where the C=C counts as two carbons. So priorities 1 and 2 are assigned to the -Br and the -OH, respectively. Determining 3 and 4 then fall into a similar pattern where we go layer by layer:he C that is labeled 3 has a carbon-carbon double bond, so it counts as C, C, H, while the carbon that is labeled 4 has C, H, H. It should be noted that in (c) the alkene is converted into a group that has single bonds by using phantom atoms, which are denoted C and do not extend beyond that.

B) Reorienting the molecule (step 2) requires thinking in 3D and redrawing the molecule. This can lead to errors when the molecule is redrawn. An alternate method, which we will dicuss here, uses your hands to guide you. To use this method, assign priorities as normal. Then look at where group 4 is. There will be three scenarios: (1) group 4 is a dash, (2) group 4 is a wedge, and (3) group 4 is in the plane (neither dash nor wedge.) We can see these possibilities in Figure 14, which are all representations of the same molecule, (1*S*)-1-bromo-1-chloroethane.

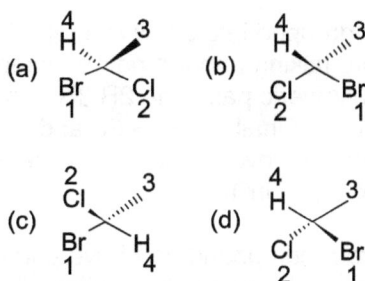

Figure 14: Four different representations of (1*S*)-1-bromo-1-chloroethane

To use this method, all you will need are your hands. The advantage of this method over the standard "steering wheel" route is that you do not need to redraw the molecule. In this method, you point your thumb toward group 4, and observe if the natural curl of *one* your hands matches the groups labeled 1,2,3 of the other priority groups. It should be noted here that 2,3,1 and 3,1,2 are also acceptable matches. Also, this will only match for one hand. Thus, for (a) you want to point your left hand in a thumbs down, where your thumb is touching the paper/screen. As you curl your hand as if you were turning a door knob, you will notice that the numbers match the curl. Thus, it is assigned *S*. Now, try it with your right hand. You will note that it does not match your right hand. In (b), group 4 is a wedge. When you align your thumb with group 4, you will make a thumbs up. If your hand is on a flat surface, it will be a thumbs up; if it is on a screen, your thumb will be pointing toward you. That is, it will be 90 degrees to the page and your thumb will be pointing *away*. If you use your left hand, once again it will match one of the trios mentioned. Your right hand will *not* match.

In (c) and (d), group 4 is in the plane of the paper (neither dash nor wedge). In this case, your hand will lay flat. You may need to rotate your hand to get your thumb pointing toward group 4. In this guided explanation, you should use your left hand palm-down, with your thumb pointing toward group 4. Your thumb should be pointing toward 4 o'clock on a clock face. Now, if you visualize the stereogenic carbon being in the center of your hand, you will realize that your fingers are *between* groups 2 and 3. As you curl your left hand, your fingers will run into group 3 first. As you continue to curl around, you will run into group 1 next, and finally group 2; 3,1,2 is OK, and thus this matches *S*. In (d), your left hand should be palm up with your thumb pointing toward 10 o'clock. Here, your fingers will be aligned with group 3 in the plane. When you curl your hands, you can choose to start with group 3. As you move around, you will next hit group 1, since it is *above* the plane, and then finally group 2. Again, this method works for any tetrahedral stereocenter.

Now that we have talked about assigning R and S to stereocenters, we need to talk about how these relate to enantiomers, diastereomers, and meso compounds. We can discuss all of these topics using compounds that have two stereocenters. When considering compounds with R and S stereocenters, the theoretical number of stereoisomers can be calculated by the following formula: 2^n, where n is the number of stereocenters, so two stereocenters will give four possible stereoisomers, shown in Figure 15 for 2-bromo-3-chlorobutane, a compound with two chiral centers.

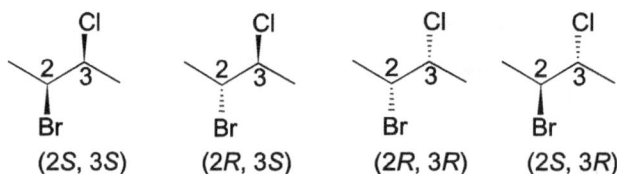

Figure 15: Possible stereoisomers of 2-bromo-3-chlorobutane

These four stereoisomers can be grouped into a few isomer relationships. Whenever we talk about enantiomers or diastereomers, we are talking about a pair of compounds and how they relate to each other. 2R,3R and 2S,3S are an enantiomeric pair. The 2R,3R and 2R,3S pair and the 2R,3R, and the 2S,3R pair are both diastereomeric pairs. Finally, the 2R,3S and the 2S,3R pair is also an enantiomeric pair. Since enantiomers are mirror images, every single chiral center must be inverted to be an enantiomer. These relationships are summed up in Table 3.

Finally, there are situations where a compound can have chiral centers, and yet the compound is achiral. These are called meso compounds, symmetrical compounds that have a mirror plane internal to the molecule. A few examples are shown in Figure 16.

It should be noted here that if the molecules can freely rotate, they should be aligned as symmetrically as possible when checking. Notice also that everything that is on one side of the mirror plane reflects exactly to the other. As we saw in Table 3, the two stereocenters generated four stereoisomers. Since each stereocenter has two 'options' (R and S), the theoretical maximum number of stereoisomers when

Table 3: Relationships between stereoisomers of 2-Bromo-3-chlorobutane

Relationship	(2S,3S)	(2R,3S)	(2R,3R)	(2S,3R)
(2S,3S)	Same	Diastereomer	Enantiomer	Diastereomer
(2R,3S)	Diastereomer	Same	Diastereomer	Enantiomer
(2R,3R)	Enantiomer	Diastereomer	Same	Diastereomer
(2S,3R)	Diastereomer	Enantiomer	Diastereomer	Same

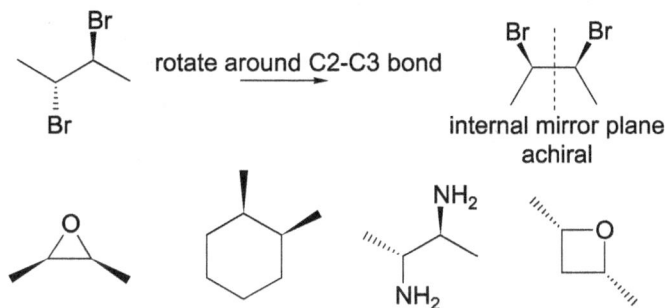

Figure 16: Examples of *meso* compounds

we have a compound with n chiral centers is 2^n. This assumes there are no *meso* compounds. Let's look at a similar case, but now we are going to study 2,3-dibromobutane. We can see the isomers in Figure 17.

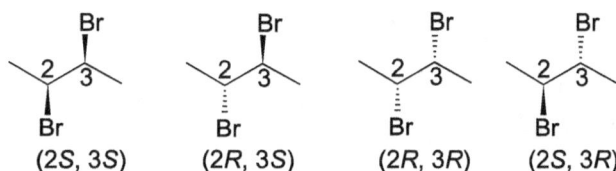

Figure 17: Potential Stereoisomers of 2,3-dibromobutane

We can see by looking at these four compounds that the two structures that have one *R* and one *S* stereocenter are actually the **same** compound! If you rotate either of them 180 degrees in the plane of the paper, it converts to the other structure. Thus, *meso* compounds serve to reduce the number of *actual* stereoisomers of a given compound. If the compound has the proper structure that can show symmetry, the meso isomer is the one that has one stereocenter *R* and one stereocenter *S*. We can see the impact this has on the isomer relationships when we look at a similar table of relationships of the stereoisomers of 2,3-dibromobutane in Table 4.

Table 4: Relationships between Stereoisomers of 2,3-Dibromoobutane

Relationship	(2S,3S)	(2R,3S)/(2S,3R) (meso)	(2R,3R)
(2S,3S)	Same	Diastereomer	Enantiomer
(2R,3S)/(2S,3R) (meso)	Diastereomer	Same	Diastereomer
(2S,3R)	Enantiomer	Diastereomer	Same

Here we notice something a bit different in that the meso isomer is a diastereomer to both of the other isomers since it is achiral. For this particular compound, there are three stereoisomers: the (2S,

$3S$) isomer, ($2R$, $3R$) isomer, and the meso compound.

Nucleophile vs. Electrophile

The term *nucleophile* means it is a molecule or ion seeking a nucleus. We often denote this by the terms or phrases Nuc or Nuc: which has a lone pair attached to it. When we are dealing with nucleophiles, there is a pair of electrons drawn moving toward a nucleus. The only exception to this is if it is a proton. In that case, it is called a base (see the section "Nucleophile vs. Base"). An electrophile is the opposite, something that seeks electrons. Thus, the arrow is often drawn *toward* our electrophile. Related concepts are nucleophilic and electrophilic. These are often used to describe how badly the nucleophile or electrophile wants the nucleus and electrons, respectively. There are some patterns that we tend to see when we have reactions of nucleophiles (Nuc) and electrophiles (Elec), which are represented in Figure 18. In this figure, we are looking immediately after the attack, not at final products.

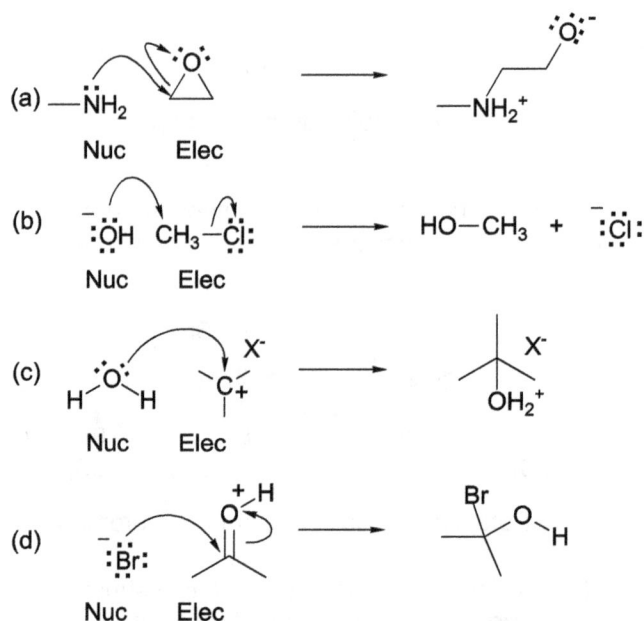

Figure 18: Four reactions of nucleophiles reacting with electrophiles

(a) Neutral nucleophile attacks neutral electrophile. Here a primary amine attacks an epoxide. After the reaction, the nitrogen is positively charged, and the oxygen is negatively charged.

(b) Negatively charged nucleophile attacks neutral electrophile. Here hydroxide attacks methyl chloride. The oxygen becomes neutral and the chlorine becomes the negatively charged chloride.

(c) Neutral nucleophile attacks positively charged electrophile. Here water attacks a carbocation. The oxygen becomes positively charged, and the carbon becomes neutral.

(d) Negatively charged nucleophile attacks positively charged electrophile. In this case, bromide attacks protonated acetone. The bromine has become neutral and the oxygen also becomes neutral.

In general, positively charged species are poor nucleophiles and negatively charged species are poor electrophiles. Thus, those options are not represented here. In all cases, the nucleophile becomes one formal charge more positive (i.e., 0 to +1). To keep the charge balanced, the electrophile will become one unit more negative (i.e., 0 to -1).

Nucleophile versus Base

The difference between a nucleophile and a base is which atom is being attacked. As we saw in the previous section, the nucleophile was attacking atoms that were not a proton. When the nucleophilic atom is attacking a proton, that is a deprotonation in the Bronsted-Lowry sense. One big example of this is comparing the S_N2 and E2 reactions. If the atom attacks a proton, we deem it a base, and we have an E2 reaction, while if the atom attacks a different atom (typically carbon) we deem it a nucleophile, and we have the S_N2 reaction. Thus, a species with a lone pair or π-bond can be a nucleophile or base, depending on what is happening around it.

There are a number of factors that go into a species' nucleophilicity, and we will look at some of the more more important ones here: charge, electronegativity, atom size, solvent, and steric hindrance. Some of these variables interact with one another, so we need to keep that in mind as we analyze. Also note that, generally, we compare two species to see which is more nucleophilic.

1. **Charge.** This one is fairly straightforward: Negatively charged species are more nucleophilic than the corresponding neutral species. Thus, a natural comparison is that water (H_2O) is a weaker nucleophile than hydroxide (OH^-) and ammonia (NH_3) is a weaker nucleophile than amide (NH_2^-). The negatively charged species is a better nucleophile because it has more electron density.

2. **Electronegativity.** Here, we are concerned with elements in the same row. Thus, we can say that the less electronegative atom in a row is the better nucleophile. For example, amide is a better nucleophile than hydroxide.

3. **Atom size.** In this case, we typically look down a column of the periodic table. For example, comparing alcohols to thiols (ROH to RSH), the thiol, containing the larger atom, is more nucleophilic. The same reasoning applies to amines vs phosphines. Again, these comparisons are best apples to apples (i.e., trialkylamines versus trialkylphosphines and not triarylamines versus trialkylphosphines for instance).

4. **Solvent.** While solvent is sometimes not specified in general reaction schemes, it can be critically important for a reaction's success for failure. In terms of nucleophilicity, solvent impacts nucleophilicity in the following way: The more solvent molecules are around the nucleophile, the less nucleophilic the nucleophile tends to be. A common impact of this is to note that for polar protic solvents (solvents that can H-bond) the nucleophilicity order for the halogens increases in the order fluoride, chloride, bromide, and iodide. Fluoride is the least nucleophilic because it is both small and basic, leading to a large number of solvent molecules surrounding the fluoride. In order to react, recall that these solvent molecules need to be removed. In polar aprotic solvents (DMSO, acetone, acetonitrile), the order is reversed, with iodide the least nucleophilic and fluoride the most. In polar aprotic solvents, there is no chance for hydrogen bonding.

When we talk about bases, we are talking about a thermodynamic characteristic. In practical terms, we consider the ratio of E2 to S_N2 to be determined by the ratios of the activation energies. This is because for each case, the energy barrier for the *reverse* reaction is high enough that under the reaction conditions, there is not a significant reverse rate. Recall that polar protic solvents make negatively charged reactants less nucleophilic. Thus, if a species such as ethoxide is used in ethanol solvent, it is much more likely to promote E2 (i.e., be more basic than nucleophilic), and if ethoxide is in a polar aprotic (e.g., DMSO), it is more likely to promote S_N2 reactions.

Two things to note here: (1) Running an actual reaction is really the only way to really see what the reality is and (2) when learning organic chemistry, instructors often will draw sharper distinctions than what happens in reality for $E1/E2/S_N1/S_N2$ reactions to make it easier to learn the material the first time. For example, an instructor may indicate that anions with a pKa of less than a particular value act as nucleophiles, and anions that are higher than that pKa act as bases. When in doubt, ask your instructor for clarification on these issues!

Resonance

Resonance is a concept that often suffers from miscommunication and misinformation. There are many many pages and videos on the internet that claim to explain the concept. Many are good; however, many are also not good. Hopefully, this primer will make the topic easier to understand.

One of the basic concepts of chemistry is that systems generally tend toward stability when possible. Thus, when we have anion, cation, or radical species that are capable of spreading charge or electron density to multiple atoms, it will do so. This occurs because when the density is spread out, the system becomes more stable. Thus, resonance may be more clearly thought of as delocalization, that is, a spreading out of charge or electron density.

Delocalization occurs when we have a charge or unpaired electron next to a π-bond. Whenever we discuss resonance or delocalization, we are talking about π-bonds and electrons and never σ-bonds. A couple things to note about resonance structures: (1) They are not real. They are convenient fictions we use to quickly predict properties and reactivity. (2) The double-headed arrows we see are specifically used to denote resonance structures. To begin, we will start with two particular systems that show up regularly in organic chemistry: allylic systems and benzylic systems. Figure 19 shows allylic and benzylic anions, cations, and radicals along with the individual resonance structures that we use to help describe what is going on.

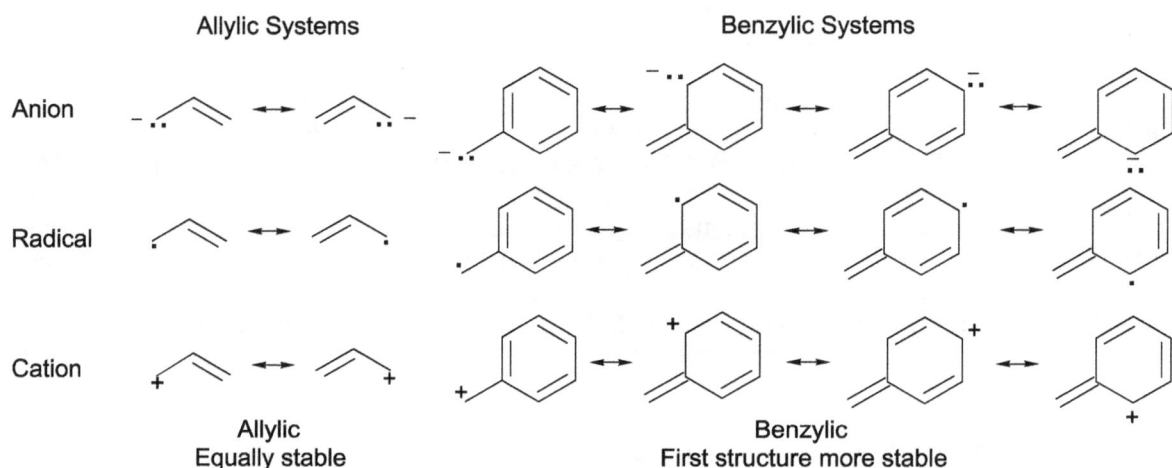

Figure 19: Resonance in allylic and benzylic anions, radicals, and cations

These are individual structures that have the charge or radical at different locations. The actual structure of these species is an average of the resonance structures, with the precise percentage mix depending on the stability of the individual resonance structure.

In the allylic systems, the density/charge is split evenly between the carbon on each end of the system. For the benzylic systems, the first resonance structure shown is much more stable than the other three. However, the other three are key to describing and predicting the reactivity of these systems, even if it is a minor contributor to the overall picture. A few other examples will help illustrate this point. First, we have acetone (see Figure 20).

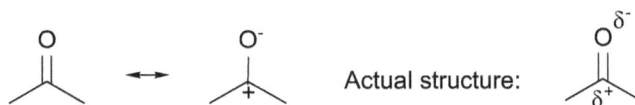

Figure 20: Resonance structures of acetone

There are two main resonance structures for acetone-one on the left of the resonance arrow where the C=O bond is intact and one on the right of the arrow where the negative charge is on the oxygen, and positive charge on the carbon. The actual structure closely resembles the structure where the C=O is intact, but the oxygen has a partial negative (shown as δ^-) charge, and the carbon a partial positive charge (shown as δ^+). Delocalization and resonance structures help predict reactivity. For acetone, we know that nucleophiles tend to attack the carbonyl carbon because it does have that partial positive charge, rendering it electrophilic. The oxygen of the carbonyl tends to attack Bronsted acids or Lewis acids since it is a partial negative charge, making it basic/nucleophilic. Next, let's look at an enolate and an enamine, which are common reactive species in synthetic organic chemistry (see Figure 21.)

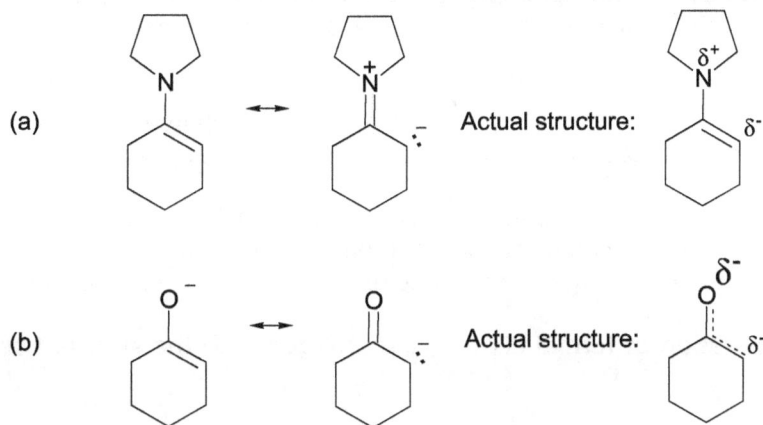

Figure 21: Relation of individual resonance structures to actual structures

Note how, for each compound, one the two resonance structures has a negative charge on the carbon. For the enolate, the structure with the negative charge on the more electronegative oxygen is much more stable than that of the structure where the negative charge is on the carbon. For the enamine in (a), the structure that has both atoms neutral is more stable than the one that has a negative charge on the carbon and a positive charge on the nitrogen. Thus the actual structure looks a lot like the first structure, but there are partial charges on the molecule. In (b), there is already a negative charge on the enolate, so both resonance structures will have a negative charge. The first resonance structure is more stable since the negative charge is on the more electronegative oxygen. For both of those species, there is a delta minus at the carbon, which indicates they can react as nucleophiles or bases at the carbon. Finally, let's look at an example with a bit of an extended π-system. Two examples can be seen in Figure 22.

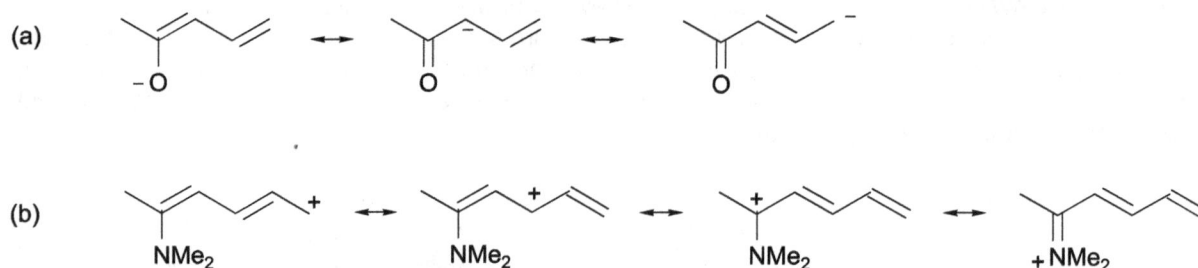

(a)

(b)

Figure 22: Resonance structures of an anion and a cation in an extended π-system

In (a), we have an enolate that is conjugated, so the negative charge can be spread out among two other carbons. As indicated by the resonance structures, that enolate can potentially react at the carbons as well as the oxygen. In (b), we have a cation that is conjugated. Notice how the plus charge moves down the chain towards the nitrogen, and then ends up *on* the nitrogen in the structure all the way to the right.

As mentioned earlier, there are more and less stable resonance structures, depending on where the charge is (i.e., negative charge on oxygen is stable compared to carbon). Here are a few tips to find the most stable resonance structures:

1. **Compounds that have no formal charges are often the most stable structure.** If we consider the acetone structures, the structure that has no formal charges is much more stable than the one next to it.

2. **Octets are preferred.** Thus, the structure that has a positive charge on an oxygen or nitrogen (the last structure in (b)), is more stable than one with a plus charge on carbon, if it gives all atoms an octet.

3. **Charges prefer to be on atoms that can best stabilize them.** If overall negatively charged, the negative charge prefers to be on an electronegative element (i.e., on nitrogen or oxygen in preference to carbon). If overall positively charged, the charge prefers to be on a less electronegative element.

4. **If there is a separation of formal charge, the charges with the smallest separation is most stable.**

In conclusion, resonance (or delocalization) can be used to predict reactivity in a number of common organic reactions. Any particular resonance structure is not accurate-a weighted average of the resonance structures comes close to showing the reality. Resonance structures are useful shortcuts to predict reactivity or properties quickly without having to generate all the details.

Interpreting Reaction Schemes

One aspect of describing organic chemistry reactions is something called a reaction scheme. While there are no exact standards for these, they generally have enough detail so that you can see what is likely to happen. You often will not be able to reproduce a reaction based on the reaction scheme alone (to do that, you often look at the experimental section in the primary literature, not in textbooks). The reaction schemes in textbooks are often designed to be part of the learning experience to emphasize particular aspects. Some examples of reaction schemes are shown in Figure 23.

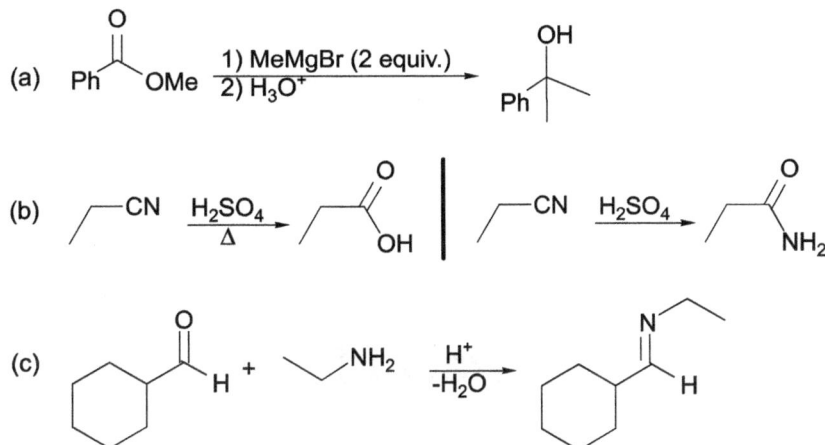

Figure 23: Examples of reaction schemes

Here, we have three different types of reactions. In (a) we have the addition of a Grignard reagent to an ester resulting in a tertiary alcohol product. Notice there are two numbered steps; this indicates that the Grignard reagent is added first, allowed to react, and then the acid is added to quench the reaction. In (b), we have a pair of similar reaction conditions. On the left is the hydrolysis of a nitrile to give a carboxylic acid. Here, we should note that there are no numbered steps, but there is the Δ symbol as part of the conditions. This indicates the reaction is heated. If this is denoted in the reaction conditions, it is likely there to give you a hint. In this case, it indicates the reaction will give the acid product. The reaction on the right does not have extra heat, so this is reaction will end as the primary amide. In (c), we have an aldehyde and a primary amine reacting under acidic conditions to give an imine. We can tell the conditions are acidic since we see (H^+) in the conditions. We also see $-H_2O$, indicating loss of water. Again, this is a hint if we see something like this included. Here, it indicates we are doing a condensation reaction.

Part III

Making the Grade

Preview of Part III

The third and final part of the book is focused on going through and looking at particular questions that you may see on organic chemistry exams. Not every possible format or question can be covered, but the questions here are solid examples to use as an aid in studying and learning the material. This section begins by having a short discussion of the two main types of questions that are generally seen; questions where you choose an answer (i.e., multiple choice, matching, and so on) where you must choose the correct answer from a group of answers, and open response, where you need to generate an answer of some type. The perception is that multiple-choice exams are easier than open response, but depending on the question and the answers given, questions where you choose an answer can be quite challenging as well. After a brief discussion of those answers, examples of questions on chosen topics will be discussed.

Types of Questions

Organic chemistry has a wide variety of topics that are covered, but the questions asked can be broken down into two main formats:

- questions where you **choose** an answer
- questions where you must **generate** an answer.

Depending on the time allowed for an exam, the topics of the exam, and the number and type of questions, being able to generate or choose an answer relatively quickly will be useful, as undergraduate exams are typically time-limited. Question formats that involve choosing an answer include multiple choice, matching, and choosing an answer. Open-response questions can be very broad in their format, but can come from any part of organic chemistry, including but not limited to: nomenclature, physical properties, spectroscopy, reactions, and synthesis. In addition, there are questions that require both types of answers (i.e., choose an answer and give why). This book is meant to be a guide for answering the most common types of questions; not all questions can be anticipated! However, the more questions you practice, the more likely you are to have seen a similar question. Having a good strategy for particular types of questions will be useful. Next there is a brief discussion of the two broad types of questions, and then we begin going through examples of questions that can come from particular topics.

Choosing an Answer

Multiple Choice

The most common type of question in this area is the multiple-choice question, with which many of us are already familiar. These questions typically have the given question followed by four or five potential answers. One of the given answers is correct, and the others are incorrect. With organic chemistry, the incorrect answers often come from common mistakes or misconceptions. This makes multiple-choice questions potentially challenging since there will be at least two answers that appear reasonable. **A good initial strategy is to generate your answer (the correct one, right?)** and compare it to the given choices. If you are confident in your answer, then the question is very simple to deal with: You just choose the same answer you generated. However, if you generate an answer and don't see it in the list, then a solid strategy will be to eliminate answers you know cannot be correct. It may be that you remember that Pd/C reduces alkynes to alkanes, so you know the answer cannot have an alkene or alkyne within it, for example. Anything you can do to reduce the number of potential answers will be useful.

Matching

The other main type of question that involves choosing is matching. For example, there may be 9 reactions without conditions, and a set of conditions that must be matched to each reaction. There may be more or less than nine choices, depending on the question. Another example is matching of names with structures, or organic chemistry terms with definitions. While these types of questions are guaranteed to have the correct answer on the page, they can span a range of difficulties and Bloom's taxonomy levels.

Generating an Answer

The other category of question is open response, which is made up of all the other types of questions that require generating a response. What is required as an answer will vary depending on the question. Some questions are relatively simple, involving memorized facts such as pKa values or bond angles or definitions. Others are very involved, such as proposing or evaluating a synthesis of a particular compound. As with the questions that involving choosing an answer, they span a range of Bloom's taxonomy levels.

Nomenclature

In organic chemistry, **there are two systems of naming for organic compounds**: the common names, and the IUPAC (International Union of Pure and Applied Chemistry) names, which are more systematic. Organic chemistry developed before systematic naming, so there are many compound names that existed previous to the IUPAC system. Many of those 'original' names are conserved in the IUPAC system. Having a systematic way of naming is very useful because scientists in one country need to have a way to communicate the structure of a particular compound to another scientist in a different country. IUPAC updates the naming recommendations from time to time, and it is possible that by the time you read this, IUPAC may have changed their recommendations slightly! This book bases its nomenclature recommendations on the recent publication in Pure and Applied Chemistry.[8] The IUPAC rules allow for multiple systematic names, depending on the structure. So when in doubt, ask your instructor about which set of rules should be used. This chapter will give practical advice and examples of nomenclature for the most common functional groups.

There are two ways nomenclature can be asked about: **directly and indirectly.** Indirect ways of asking about nomenclature are to ask about a particular named compound without giving the structure. For example, a question may ask, "Starting from bromobenzene, propose a synthesis of benzoic acid." In addition to any reactions needed, you will need to know (or derive) the structure of bromobenzene and benzoic acid. These embedded questions can show up in any part of organic chemistry. **Direct questions fall into two general types** (see Figure 1.)

Q: Draw 2-chlorobutane   Q: Name this compound:

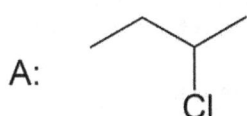

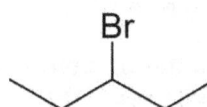

A: A: 3-bromopentane

Figure 1: Direct nomenclature questions

- **Name to structure** (i.e., draw 2-chlorobutane, or which of the structures below is 3-chlorohexane, etc.) This involves taking the name (systematic or IUPAC) and converting to a structure.

- **Structure to name** (i.e., provide a name for a particular structure, or which of these structures is 2-chlorobutane). This involves applying the IUPAC rules to generate a systematic name.

[8]Hellwich, K-H.; Hartshorn, R. M.; Yerin, A.; Damhus, T.; Hutton, A.T. Brief Guide to the Nomenclature of Organic Chemistry *Pure Appl. Chem.* **2020**, *92*, 527-539.

The formats for these direct questions may be multiple choice, open response, matching, and so forth. but they all involve one of the two types shown. The examples in this chapter will be a mixture of both types. When creating a systematic name, IUPAC has a number of steps to be taken; we will go through each example functional group separately. Large molecules can have systematic names that can take up whole lines of text, so we will stick to smaller molecules to get the hang of naming compounds.

Since many organic compounds involve alkane chains, knowing some carbon prefixes will be useful. Table 1 lists the name of the alkane, the name as a substituent, and the number of carbons. The name is that of the non-branched isomer, while the substituent column is for when the compound (minus a hydrogen) is used as a substituent.

Table 1: Nomenclature for Alkanes up to 10 Carbons

Number	Name	Substituent
1	methane	methyl
2	ethane	ethyl
3	propane	propyl
4	butane	butyl
5	pentane	pentyl
6	hexane	hexyl
7	heptane	heptyl
8	octane	octyl
9	nonane	nonyl
10	decane	decyl

The rest of the chapter is sorted by functional group, and the functional groups in the molecule are important because there is a ranking of groups that will determine how the compound is named. That is, some compounds are named as a ketone, as an amine, carboxylic acid, and so forth. Each section will have only compounds named as a member of that group. The IUPAC publication mentioned earlier has an ordered list of these characteristic groups. The group highest on the list determines how the compound is named and ends up at the end of the name (suffix). Other functional groups are listed as prefixes. For each section, if that group is on the list, the suffix and prefix will be noted. When generating a name for a compound, alkanes, alkenes, and alkynes are not considered functional groups.

IUPAC also has a general method of generating names. They are paraphrased here (in no particular order), and we will refer to them when needed.

- What is the compound named as? That is, **what is the characteristic functional group in the molecule?** IUPAC has a list of these and their ranking, but the examples we see will have only one of these groups.

- Where applicable, **find the base hydrocarbon**. For many groups, this will be the longest chain.

- **Number your carbons** in a manner consistent with IUPAC guidelines. This will also depend on the characteristic functional group. This breaks down into two types of numbering: (1) Where the lowest substituent numbers can be generated and (2) where the numbering starts from a carbon on or near the characteristic group. Which method to use will be apparent from the examples.

- **Identify your substituents**, and note where they are located. The number of the carbon they are on is called the **locant**.

- **List your substituents in alphabetical order**, grouping them together if there are two or more of them (di, tri, etc.). These prefixes do *not* impact the ordering.

- **Add any stereochemical descriptors** (E/Z or R/S) next to the locant, and write the name out.

This may seem complex, but don't worry, there are plenty of examples and practice problems in this section and at the end of the section. For the examples of alkanes, alkenes, and alkynes, all the carbons will be numbered, while for other functional group sections carbon 1 and any locant in the name will be labeled to keep the compounds readable. For most of the names we will encounter, we will use the substitutive system mentioned. For particular functional groups, such as esters, functional class nomenclature is preferred, and for several others (ketones, ethers, and sulfides), this nomenclature is still used. Examples of those will be noted as they come up.

Alkanes and Alkyl Halides

We will begin with alkanes, as they are often the first functional group that is mentioned in organic chemistry. As mentioned previously, this is *not* a characteristic group for nomenclature. There are two main types of alkanes, acyclic alkanes and cyclic alkanes. The difference of whether the molecule is in a ring or not is important, since the method of naming will be different for each type. This difference also impacts naming in that rings rank higher than chains when they clash. An example of acylic and cyclic alkanes is shown in Figure 2.

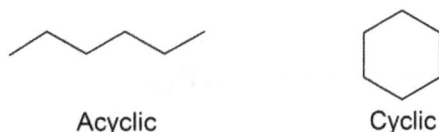

Acyclic   Cyclic

Figure 2: Examples of acylic and cyclic alkanes

The examples in this section will involve compounds that have carbons, hydrogens, and halogens. Thus, this will encompass both alkanes and alkyl halides. For acyclic alkanes, there will be two ends that we can count from. The correct procedure for alkanes is to find the longest chain, which in this case is five carbons, so the compound will be named as a pentane. This means that the base name is a pentane, *not* that the compound has only five carbons. The acyclic alkanes will have two ends to begin from regarding numbering. The correct way of numbering is to number such that the set of numbers has the lower number at the point of first difference. We can see an example of both incorrect and correct numbering of a hydrocarbon in Figure 3.

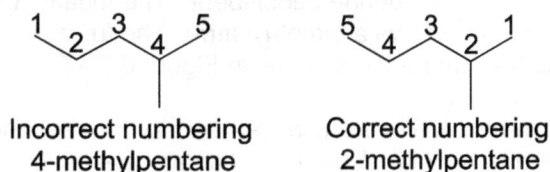

Incorrect numbering   Correct numbering
4-methylpentane   2-methylpentane

Figure 3: Example of incorrect and correct numbering

On the left, the *locant*, that is, the number that identifies where a group is, is 4, giving the name 4-methylpentane. On the right, we see the correct numbering, which gives a locant of 2, giving 2-methylpentane. As we go through more and more examples, this will get easier to do, but keep in mind both options need to be considered to find the correct way of numbering. We will now go through a number of other examples, building in complexity to show the two steps for alkanes: finding the longest, or parent chain, and providing numbers. There are a few other considerations we will also see as we go through the examples.

The next example has two substituents, a chloro group and a methyl group. The longest chain is seven carbons, so this will be named as a heptane. The numbering choices for this compound give a 4,5 number set and a 3,4 number set. As discussed previously, the numbering that gives the lower number at point of first difference is used. We can see the results of the correct and incorrect numbering in Figure 4.

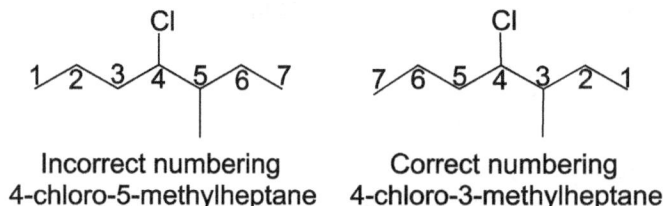

Incorrect numbering   Correct numbering
4-chloro-5-methylheptane   4-chloro-3-methylheptane

Figure 4: Example of incorrect and correct numbering

In the example with correct numbering, the substituent with the locant 4 is listed *before* the substituent with locant 3. This is because the IUPAC guidelines state that the subsituents be listed with in alphabetical order, generally ignoring prefixes such as di or tri, which indicate there are two or three of that particular group.

We have a molecule with a little more complexity in Figure 5.

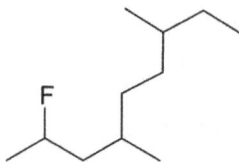

Figure 5: 2-fluoro-4,7-dimethylnonane

We still need to find the longest chain, and the chain doesn't have to be straight across. Here our longest chain will be nine atoms, so this molecule is named as a nonane. We will begin numbering from the end closer to the fluorine as that will give us the lower number at point of first difference. So here we have two methyls and one fluorine substituent. The locants of the two methyl groups are listed numerically and are combined to give a dimethyl term. The full name ends up being 2-fluoro-4,7-dimethylnonane. You can see the numbering scheme in Figure 6.

The next example is shown below. Here again we need to find the longest chain, and the longest chain in this case is eight carbons, so this will be named as an octane. The numbering starts at the end closer to the bromine, since that will give a 3-bromo instead of a 3-methyl. Here, the tiebreaker if the number locants are the same is the first letter of the substituent, here b versus m. Correct and incorrect numbering is shown in Figure 7.

The next example has one more carbon than the previous one, but the name will look quite a bit different. We can see the numbering and name in Figure 8. The main chain is now nine carbons, and

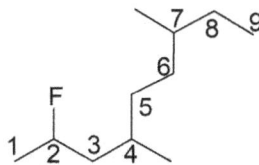

Figure 6: Numbering for 2-fluoro-4,7-dimethylnonane

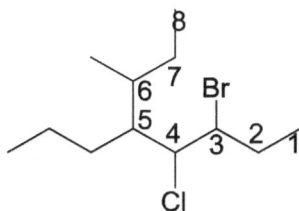

Correct numbering
3-bromo-4-chloro-6-methyl-5-propyloctane

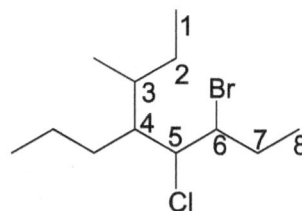

Incorrect numbering
6-bromo-5-chloro-3-methyl-4-propyloctane

Figure 7: Correct and incorrect numbering for 3-bromo-4-chloro-6-methyl-5-propyloctane

we have a four-carbon substituent on carbon 5. The substituent is named butan-2-yl because it is four carbons long and attached to the main chain on the second carbon. Once again, it is worth mentioning that the substituents are listed alphabetically and not by locant.

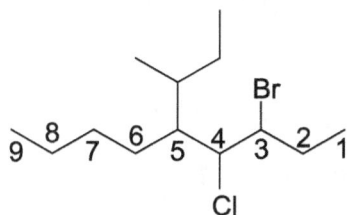

Correct numbering
3-bromo-5-(butan-2-yl)-4-chlorononane

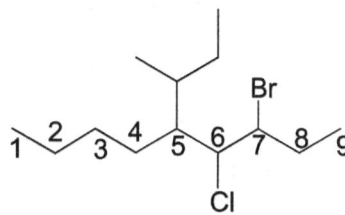

Incorrect numbering
7-bromo-5-(butan-2-yl)-6-chlorononane

Figure 8: Correct and incorrect numbering for 3-bromo-5-(butane-2-yl)-4-chlorononane

We will end the section with four more examples, these can be seen in Figure 9. Take a few minutes and try to name them yourself before continuing. Keep in mind the steps: (1) Find the longest chain, (2) number the substituents along the chain such that the lowest set of numbers is generated, and (3) write the substituents in alphabetical order, along with locants in the proper location. If the same substituent shows up multiple times, be sure to group them together and use the proper prefix. It is OK if you don't get all of them exactly right; nomenclature takes time to master.

Structure 1 has seven carbons in the chain, and the substituents are the same distance from the ends, so in either case, the numbers will be 3 and 5. As mentioned earlier, the tie breaker is which subsituent comes earlier in the alphabet, so the proper name for structure 1 is 3-bromo-5-fluoroheptane, not 5-bromo-3-fluoroheptane. Example 2 has six carbons in the longest chain, and the numbering starts at the end closer to the chlorine to give the numbers 2,3, and 4 for the substituents. The proper name for example 2 is 2-chloro-3-ethyl-4,4-dimethylhexane. Note here that the two methyl groups are on the

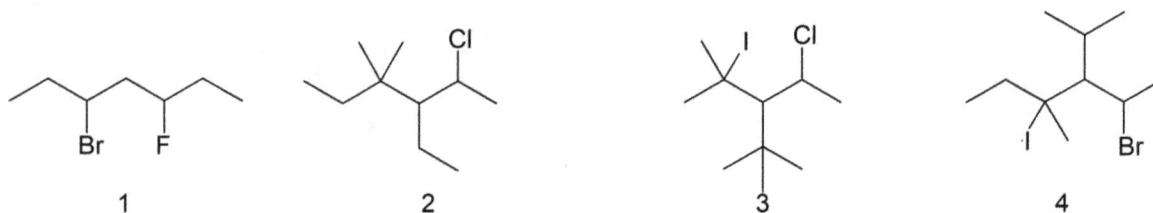

Figure 9: Four practice structures to name

same carbon, so that part of the name ends up being 4,4-dimethyl. This indicates that both methyl groups are on carbon 4. The longest chain in example 3 is five carbons, and the numbering starts at the end close to the iodo substituent and goes through the route with the two methyl groups. Thus, the name of this compound is 3-(1-chloroethyl)-2-iodo-2,4,4-trimethylpentane. For example 4, the longest chain is six carbons, and the numbering starts at the end closer to the bromine, so the name for this compound would be 2-bromo-4-iodo-4-methyl-3-(propan-2-yl)hexane. The names and numbering for these examples are shown in Figure 10.

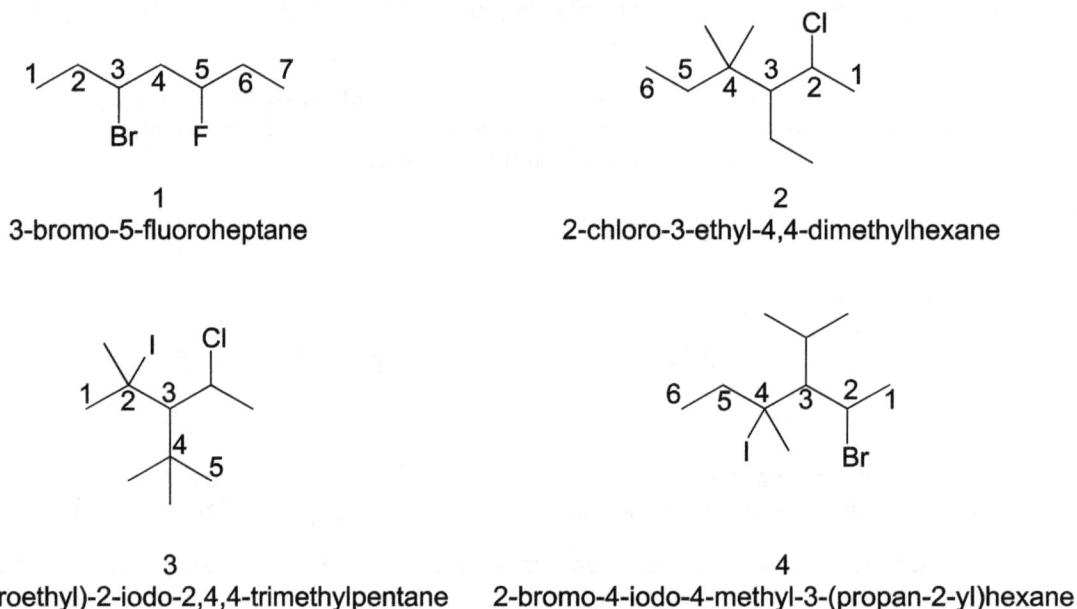

1
3-bromo-5-fluoroheptane

2
2-chloro-3-ethyl-4,4-dimethylhexane

3
3-(1-chloroethyl)-2-iodo-2,4,4-trimethylpentane

4
2-bromo-4-iodo-4-methyl-3-(propan-2-yl)hexane

Figure 10: The four practice structures named

In conclusion, we have seen a number of examples of compounds names as alkanes. IUPAC has a hierarchy of functional groups, and alkanes are at bottom. The other sections all focus on particular functional groups, and that is how a number of organic chemistry nomenclature questions are asked. The general ideas of finding the longest chain and having the lowest set of numbers carries through to a lot of other sections in this chapter.

Alkenes

Alkenes are compounds that have one or more carbon-carbon double bonds. This section will focus on alkenes and cycloalkenes that have alkyl groups and halogens as substituents. Alkenes do not count as characteristic groups. Many organic books use older IUPAC guidelines (1973 or 1993, check with your instructor for which ones they want you to use) that had the longest chain include the alkene; the new guidelines are slightly different because the longest chain does not have to include the double bond. The other guidelines we saw with alkanes are seen here as well. In Figure 11, we see a simple alkene. The only 'functional' group in the molecule is the alkene, so we need to begin numbering at that end, where the first encountered carbon in the alkene has the lower number. Here is an example of the difference between the newer and older IUPAC conventions. The older guidelines would name this as 1-pentene, where it is understood the one refers to the alkene. In the new convention, the 1 is inserted into the alkene name to remove any confusion. We will follow the new convention in this book.

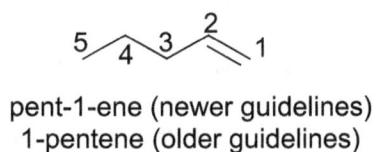

pent-1-ene (newer guidelines)
1-pentene (older guidelines)

Figure 11: Newer versus older IUPAC nomenclature guidelines

In the next example shown in Figure 12, we have an alkene that has stereochemistry. In this case, it is a Z-alkene since carbon 1 and carbon 4 are on the same side of the alkene, as noted in the stereochemistry and isomers section. Also note that the stereochemical descriptor goes with the *lower* number of the double bond. Finally, numbering the opposite way would give the alkene a locant of 3.

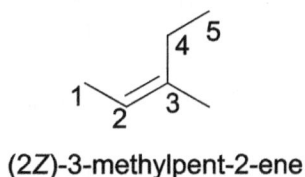

(2Z)-3-methylpent-2-ene

Figure 12: Naming an alkene that has a defined stereochemistry

In the next example, we have an example of what happens when the alkene is named as a substituent of the main chain, instead of being part of the main chain as in the first two examples.

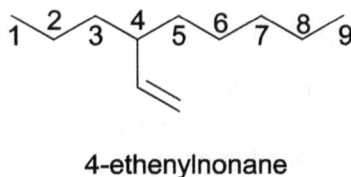

4-ethenylnonane

Figure 13: Nomenclature with alkene as a substituent

Here, in Figure 13, we see we have nine carbons, and we begin numbering closer to the end with the alkene. Again, this is different than in previous IUPAC guidelines, since the name would be 3-propyloct-1-ene if the old system was used. Instead we have 4-ethenylnonane. The substituent is ethenyl since the IUPAC name for the two carbon substituent is ethene.

Our next example, shown in Figure 14, is a diene that has one of the double bonds that has stereochemistry.

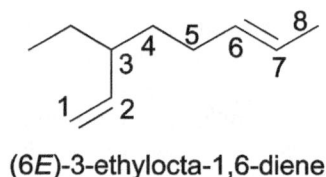

(6*E*)-3-ethylocta-1,6-diene

Figure 14: Example of diene that has one bond with defined stereochemistry

Again, we look for the longest chain. Here, both dienes are in the numbered chain because the double bond takes precedence over the ethyl group since both branches at carbon 3 contain two carbons. Since we have a terminal alkene (an alkene at the terminus of a chain or branch), we begin our numbering there to give one of the alkenes the number 1. At carbon 3 we have an ethyl group, and at carbon 6 (again, the first carbon of the double bond containing carbons 6 and 7) we see an *E*-double bond. Also note the diene contains the locants for both alkenes.

Figure 15 shows our next example.

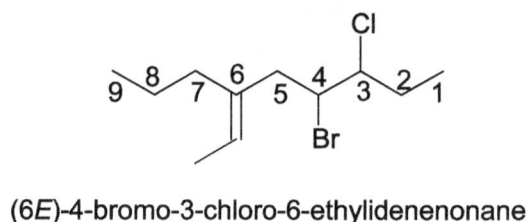

(6*E*)-4-bromo-3-chloro-6-ethylidenenonane

Figure 15: Example of alkene that has one of the double-bond carbons in the main chain

Here we have a couple of things to note. First, the longest chain again does *not* contain the alkene. Second, one of the alkene double-bond carbons is part of the chain. The alkene substituent is named ethylidene because it has two carbons and has one of the double bond carbons *in* the main chain. The numbering begins closer to the chlorine substituent since it has the number 3, whereas starting at the other end would give a number of 4. Again, note that the substituents are listed in alphabetical order.

Next, we have a few examples of cyclic alkenes. The first is shown in Figure 16.

Here, there are a few things of note. First, the alkene is part of the ring, so the numbering will begin at one of those two carbons. Since one of the alkenes has a methyl group, that is where the numbering will begin. Since it is an alkene, the numbering must continue through the alkene, even though there could be lower numbers if you did *not* go through the alkene with numbering (i.e., the chlorines could be 3 and 4 if you went counterclockwise). Since there are two chlorines, the name contains dichloro.

Our next example is shown in Figure 17.

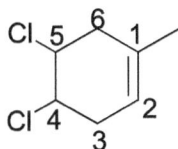

4,5-dichloro-1-methylcyclohex-1-ene

Figure 16: Example of a cyclic alkene

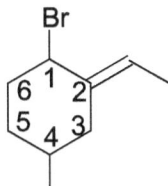

(2*E*)-1-bromo-4-methyl-2-propylidenecyclohexane

Figure 17: Example of a cyclohexane with an alkene substituent

In contrast to the previous example, the alkene is external to the ring, so it will be named as a substituent. Here, we are generating the lowest set of numbers, so our numbers will be 1, 2, and 4. The substituent on carbon 2 has three carbons, so it is named propylidene.

The next example is shown in Figure 18.

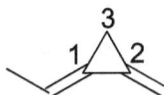

(1*E*)-1-ethylidene-2-methylidenecyclopropane

Figure 18: Example of a cyclopropane ring with two alkene substituents

Here we have a cyclopropane ring with two alkene substituents. The number 1 goes with the ethylidene since it has two carbons versus the methylidene that has one carbon. It should also be noted this is a diene but doesn't have diene in the name.

The final example in the section is shown in Figure 19.

Here, we once again have a cycloalkene, now a cyclopentene to be precise. The two carbons of the alkene have no substituents on them, so the numbering can start at the carbon that would lead to the lowest locant number. In this case, the methyl is 3. If the other carbon of the alkene was used, the lowest number would be 4.

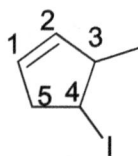

4-iodo-3-methylcyclopent-1-ene

Figure 19: Example of a cyclic alkene

Alkynes

Next up on the list are alkynes, which are compounds that contain a carbon-carbon triple bond. This is *not* a characteristic group. The numbering procedure is similar to that of alkenes, except that there are no isomers with alkynes. There are two types of alkynes, terminal and internal. Terminal alkynes have the alkyne at the end of a chain, while internal alkynes are not at the end of a chain. Another wrinkle with alkynes is that due to their sp-hybridization, rings containing alkynes tend to be eight carbons or larger. With the reduced variety of structures that alkynes can have, there won't be as many examples in this section. Our first example is a simple linear alkyne as shown in Figure 20. As with alkenes, the numbering will start on the end closer to (or containing) the alkyne. In this case, the compound is named hex-2-yne. This is an example of an internal alkyne.

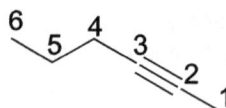

hex-2-yne

Figure 20: Nomenclature example for an internal alkyne

The next example is shown in Figure 21. This is an example of a terminal alkyne, so the numbering starts at the terminal carbon. A chlorine on carbon 6 makes the name 6-chloroct-1-yne.

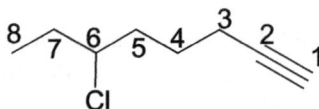

6-chloroct-1-yne

Figure 21: Nomenclature example for a terminal alkyne

Figure 22 shows a cycloalkane with an alkynyl substituent. Regardless of which way the numbering starts, the locants for this compound will be 1 and 3. Thus, the tiebreaker becomes which substituent comes earlier in the alphabet. The B in bromo ends up being before the E in ethynyl. Thus, the name is 1-bromo-3-ethynylcyclohexane.

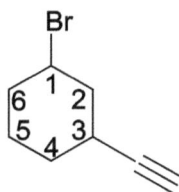

1-bromo-3-ethynylcyclohexane

Figure 22: Nomenclature example of a cyclohexane with an alkynyl substituent

Our next example has alkyne and an alkene, something that can be classified as an enyne. As we can see in Figure 23, we start on the end near the alkyne because it gives a lower locant (2) than if we start at the end closer to the double bond. If that were to occur, our locants would be 3, 6, 7, giving a name of (3E)-6-bromonon-3-en-7-yne. The correct name is (6E)-4-bromonon-6-en-2-yne.

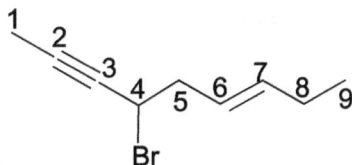

(6E)-4-bromonon-6-en-2-yne

Figure 23: Nomenclature example of an enyne

Finally, we see an example of a cycloalkyne in Figure 24. Carbon 1 needs to be one of the alkyne carbons. Once again, we have a numbering tie, with the bromomethyl getting the lower number in this case 4. Thus, the name is 4-(bromomethyl)-7-chlorocyclooct-1-yne.

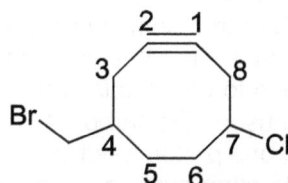

4-(bromomethyl)-7-chlorocyclooct-1-yne

Figure 24: Nomenclature example of a cycloalkyne

Aromatics

Benzene is the prototypical aromatic compound, so we will look at benzene and its derivatives in this section. However, it is *not* a characteristic group. There are a number of substituted benzenes that have common names, and those are shown in Figure 25.

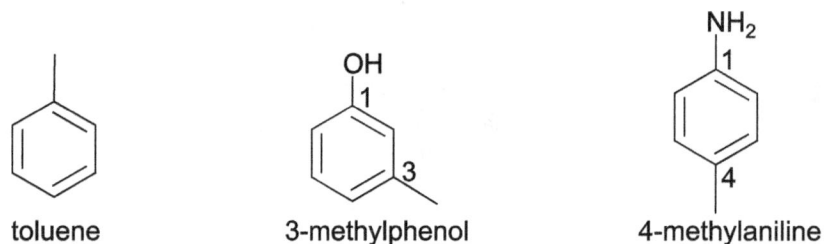

Figure 25: Nomenclature examples using common aromatic system names

Three aromatic molecules that have common names are toluene (instead of methylbenzene), phenol (instead of hydroxybenzne), and aniline (instead of aminobenzene). In many cases, once other groups are attached to toluene, the name is written as if it were a methylbenzene. One note for substituted phenols and anilines: The atom with the hydroxy group and the amino group are always numbered 1 because they are named as a phenol or an aniline. This is why we see the compounds named 3-methylphenol and 4-methylaniline.

Two examples of aromatic nomenclature are in Figure 26.

1-chloro-3-(propan-2-yl)benzene 2-bromo-4-(chloromethyl)-1-methylbenzene

Figure 26: Nomenclature examples of di- and trisubstituted benzenes

On the left, we have a 1,3-disubstituted (*meta*)-disubstituted benzene. While the *ortho, meta, para* system is acceptable for disubstituted benzene rings, you can always use numbers. Either way, the numbers will be 1 and 3, so the substituent that comes earlier in the alphabet gets the number 1. Another note about the substituent on carbon 3: This is an isopropyl or 2-propyl group. This means that a three-carbon group is attached and bound to the middle carbon. The example on the right is a trisubstituted benzene, and numbers must be used. On carbon 4 is a $-CH_2Cl$ group. This is a one carbon group, so the root is methyl. In addition, one proton has been replaced with a chlorine, so the name is chloromethyl for this substituent. As with other cyclic systems, the lowest number at first difference is preferred, so we get the name 2-bromo-4-(chloromethyl)-1-methylbenzene. The second pair of examples is shown in Figure 27.

On the left is a fairly straightforward example of 1-chloro-2-nitrobenzene. Again, the carbon with the chlorine gets to be on carbon 1 since it appears before N in the alphabet. The second example is a bit more interesting since there are multiple ways to number these as shown in Table 2.

As we can see, there are six different combinations that can be imagined for numbering the trisubstituted benzene. As with other situations with multiple potential combinations, this one breaks down by the first letter of the substituent, so we have 1-bromo-3-fluoro-5-methylbenzene.

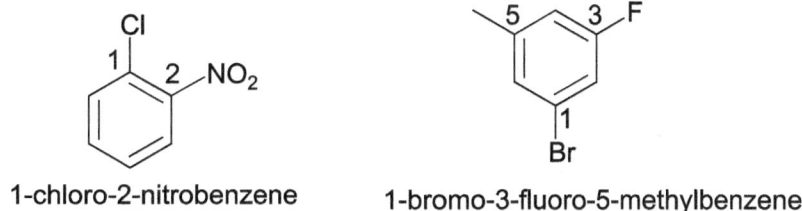

1-chloro-2-nitrobenzene   1-bromo-3-fluoro-5-methylbenzene

Figure 27: Nomenclature examples of di- and trisubstituted benzenes

Table 2: Potential Numbering Schemes for 1-Bromo-3-fluoro-5-methylbenzene

Bromo	Fluoro	Methyl
1	3	5
1	5	3
3	1	5
3	5	1
5	1	3
5	3	1

Alcohols and Phenols

With alcohols and phenols, we are moving to compounds that now have a characteristic functional group. We have seen some phenols in the section on aromatics, and there are a few more examples here. For alcohols, the suffix is -ol. This means that the name of the alkane has the e removed and -ol added, so ethane becomes ethanol, propane to propanol, and so on. As a reminder, if a compound is named as a phenol, the carbon with the -OH group gets the number one. We can see a group of three substituted phenols in Figure 28.

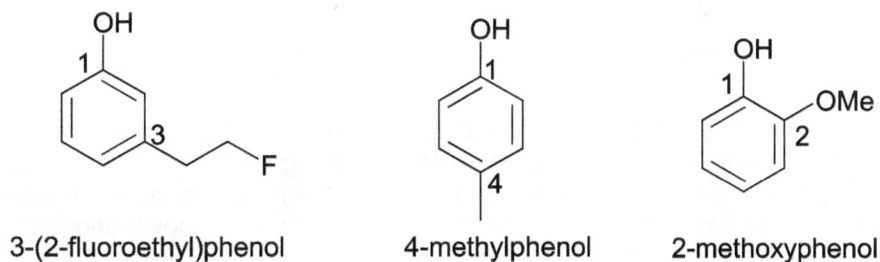

3-(2-fluoroethyl)phenol   4-methylphenol   2-methoxyphenol

Figure 28: Nomenclature examples involving phenols

On the left, we see a substituted phenol, and carbon 3 has a substituted ethyl group. The locant 2 in the 2-fluoroethyl group is there because we count the carbons on the fluoroethyl substituent from where it attaches to the aromatic ring. The middle structure has a methyl group on carbon 4, and the structure on the right has a methoxy group on carbon 2. The substituent is called methoxy because if you break that bond and add a carbon it would become methanol. If there is an ethyl group instead, it becomes ethoxy, and so on. We will see this type of nomenclature quite a bit in the section on ethers, since it turns out 2-methoxyphenol also contains an ether.

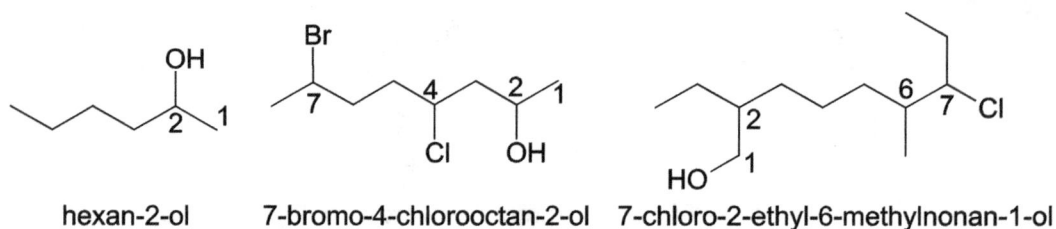

hexan-2-ol   7-bromo-4-chlorooctan-2-ol   7-chloro-2-ethyl-6-methylnonan-1-ol

Figure 29: Nomenclature examples involving alcoohols

Figure 29 has a set of three acyclic alcohols to analyze. The compound on the left is straightforward and is named hexan-2-ol. The older nomenclature recommendations would have the name be 2-hexanol. The middle example has a number of constituents, with the bromine and the alcohol both on carbons adjacent to the terminal carbons on the chain. If you think about the rules we have talked about thus far, we would predict the name to have the bromine with a locant of 2, since bromo would come before hydroxy. However, since alcohols are a characteristic functional group, the -OH will get the lower number. Thus we have the name 7-bromo-4-chlorooctan-2-ol. The final example has the -OH get the number 1, and so the name is 7-chloro-2-ethyl-6-methylnonan-1-ol. As always, keep in mind that the substituents are listed alphabetically, not numerically.

The final three examples in the section are shown in Figure 30.

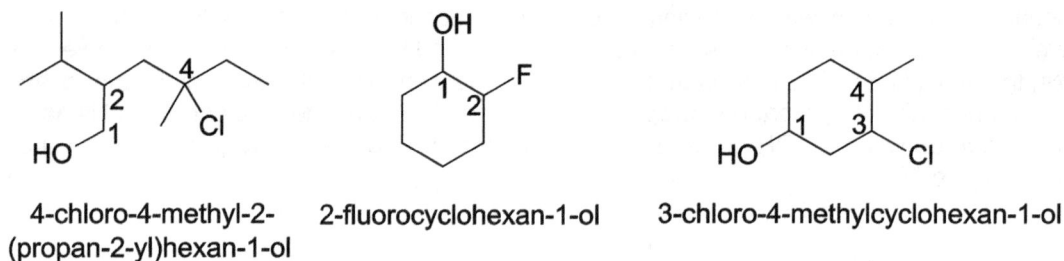

4-chloro-4-methyl-2-   2-fluorocyclohexan-1-ol   3-chloro-4-methylcyclohexan-1-ol
(propan-2-yl)hexan-1-ol

Figure 30: More nomenclature examples involving alcohols

The example on the left is interesting because it is an example where the parent chain is *not* the longest chain in the molecule. There is a seven-carbon chain that goes horizontally across the molecule. Once again, the fact that the molecule is an alcohol sets up the molecule to be named as a hexanol. The substituent on carbon 2 is a propan-2-yl group (common name isopropyl), and carbon 4 has both a methyl and a chlorine attached. Thus, the name is 4-chloro-4-methyl-2-(propan-2-yl)hexan-1-ol. The middle example is a cyclohexanol, so carbon gets the number 1. The final example again has the carbon with the -OH as 1, with the other two getting the lower numbers.

Ethers

For ethers, there are two types of nomenclature commonly used. There is the IUPAC and the common. The common nomenclature (used for simple ethers) involves putting the two groups on either side of the oxygen atom in alphabetical order and then putting ether at the end. Ethers also do not count as characteristic groups. Figure 31 has three examples of simple ethers that are named in both manners.

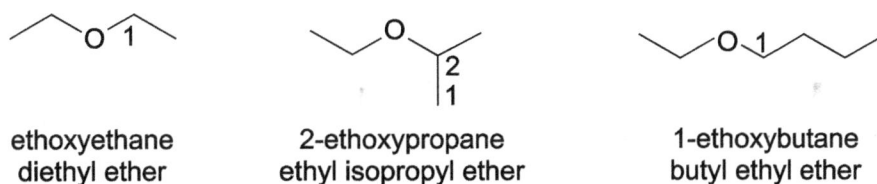

ethoxyethane
diethyl ether

2-ethoxypropane
ethyl isopropyl ether

1-ethoxybutane
butyl ethyl ether

Figure 31: Nomenclature examples involving ethers

The example on the left is named ethoxyethane. The name implies it is a subsituted ethane, and the ethoxy comes from the fact that the -OEt group is a substituent. Here, ethanol becomes ethoxy. The other way of naming generates diethyl ether, since there are two ethyl groups, one on either side of the oxygen atom. The middle structure is named 2-ethoxypropane. The locant 2 indicates that the ethoxy group is on the central carbon. The common name is ethyl isopropyl ether since one side has an ethyl group and the other has an isopropyl group. The example on the right is named 1-ethoxybutane since the ethoxy group is attached to carbon 1.

The next three are a little more complex and will have only substitutive names given. We can see those in Figure 32.

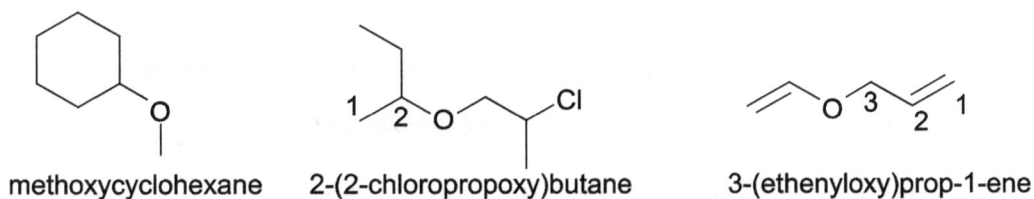

methoxycyclohexane   2-(2-chloropropoxy)butane   3-(ethenyloxy)prop-1-ene

Figure 32: Nomenclature examples of ethers using substitutive names

The structure on the left is named methoxycyclohexane because rings are ranked above chains. Regardless of the size of the ring or number of carbons in a chain, the ring is the base name for the compound. The middle structure has a three-carbon chain on the right side, and a four-carbon chain on the left side. Thus, the compound is named as a substituted butane. The -OR group is named 2-chloropropoxy since the chlorine is on the second atom, counting from the oxygen outward. The third example has a similar rationale to the middle one; there the side with three carbons is ranked above the one with two carbons. Thus, it is named as a propene. The other side is named ethenyloxy because it would be derived from ethenol.

The final set of four shows examples of nomenclature for cyclic ethers and can be seen in Figure 33.

The nomenclature system for cyclic ethers involves the prefix ox- followed by -irane for three-membered rings, -etane for four memmbered rings: -ane for five membered rings, and -olane for six membered rings. Actually it is oxa-, but since all of the suffixes have vowels, the a is removed and practically speaking for these systems, it is ox-. For all of these rings that contain one oxygen and the rest carbon, the oxygen atom gets the number 1, and the rules are followed as seen previously, giving the lowest locants possible at first difference. Of note is the third example, where the fluoro is 3 because F comes before M in the alphabet.

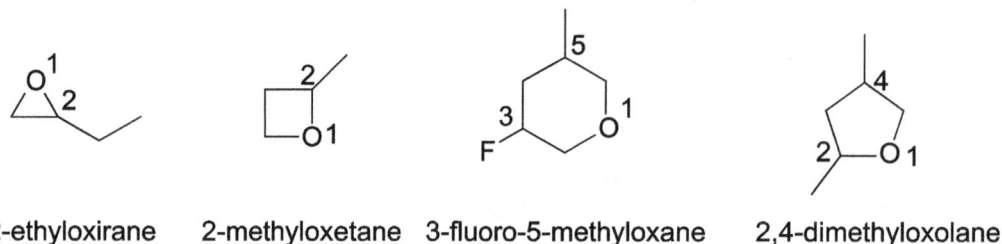

2-ethyloxirane   2-methyloxetane 3-fluoro-5-methyloxane   2,4-dimethyloxolane

Figure 33: Nomenclature examples of cyclic ethers

Thiols and Sulfides

Thiols and sulfides are named in a similar fashion to alcohols and ethers. Also similar to alcohols and ethers, thiols are a characteristic group (thiol is added to the end of the alkane name). First, we see three examples of thiols in Figure 34.

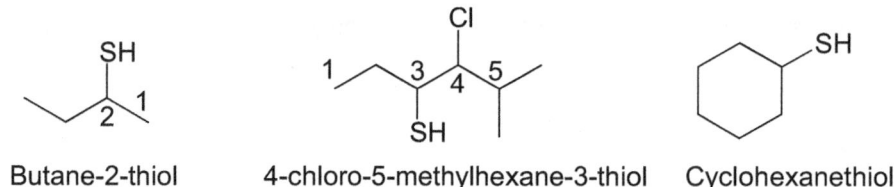

Butane-2-thiol   4-chloro-5-methylhexane-3-thiol   Cyclohexanethiol

Figure 34: Nomenclature examples of thiols

In the first example, we have butane-2-thiol. The corresponding alcohol would be butan-2-ol. The middle example is another example where we now have a 'characteristic' group, in this case the thiol. The numbering begins on the left side of the molecule so that the thiol can get the number 3 (versus 4 if started from the other side). The final example is similar to the alcohol cyclohexanol. Our next set of examples are sulfides, which are sulfur analogs to ethers. These are seen in Figure 35.

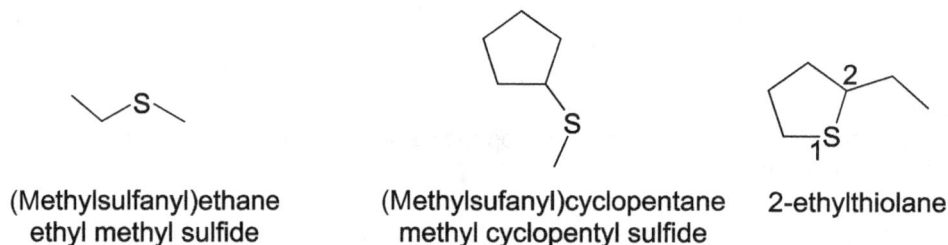

(Methylsulfanyl)ethane   (Methylsufanyl)cyclopentane   2-ethylthiolane
ethyl methyl sulfide   methyl cyclopentyl sulfide

Figure 35: Nomenclature examples of sulfides

Here we see a similar nomenclature for simpler sulfides as we saw for ethers. The first example is (methylsulfanyl)ethane, or ethyl methyl sulfide. The substitutive nomenclature is a bit different for sulfur, since we have methylsulfanyl instead of methoxy. The second example is named as a cyclopentane because rings are ranked higher than chains. The suffixes for the rings are the same as we saw for cyclic ethers. The prefix used for these systems is thi-. Actually it is thia-, but since all of the suffixes have vowels, the a is removed and practically speaking for these systems, it is thi-. The suffixes for ring sizes 3 to 6 are -irane for three membered rings, -etane for four memmbered rings, -ane for five membered rings, and -olane for six-membered rings. Thus, the third compound is named as 2-ethylthiolane.

The final set of examples contains a slightly more complex sulfide in Figure 36.

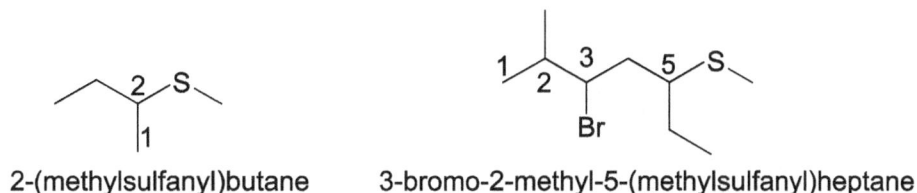

2-(methylsulfanyl)butane   3-bromo-2-methyl-5-(methylsulfanyl)heptane

Figure 36: Other nomenclature examples of sulides using substitutive nomenclature

The first example is named as a butane since the left side has four carbons, and the right side has one carbon. The second example is named as a heptane, since the longest carbon chain is seven carbons. We start numbering on the left side, since we get a lowest locant of 2, whereas if we start on the end closer to the sulfur, our name would be 5-bromo-6-methyl-3-(methylsulfanyl)heptane.

Aldehydes and Ketones

Aldehydes and ketones are both carbonyl-containing functional groups. The aldehyde has a proton and an R group attached on either side of the carbonyl, while aldehydes have a proton and an R group on either side of the carbonyl. We see our first set of aldehydes in Figure 37. Both aldehydes and ketones are characteristic functional groups that have a suffix. For aldehydes, there are two different ways the suffix is used. The first is if the aldehyde is at the end of a chain. In this case, the carbon is implied in the name, and the -ane of the alkane is replaced with -al, so propane becomes propanal. The other case is when the aldehyde group is a substituent, in particular on a ring. Thus, an aldehyde attached to a cyclohexane ring would get the name cyclohexane-1-carbaldehyde. For ketones, the position is noted with a locant, and the -e of the alkane is replaced with -one, so a three carbon ketone becomes propan-2-one, and so on.

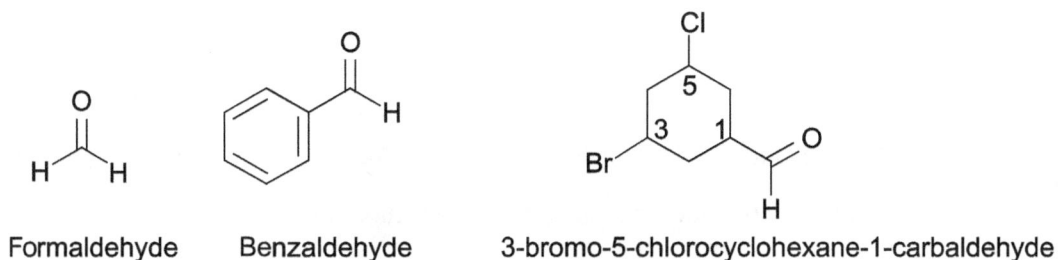

Formaldehyde   Benzaldehyde   3-bromo-5-chlorocyclohexane-1-carbaldehyde

Figure 37: Nomenclature of aldehydes

We can see a couple of common aldehydes in the first two structures. The first is a unique aldehyde called formaldehyde. It is the only compound that has a proton on either side of the carbonyl. The second is benzaldehyde, which is a very common compound. The third example is an example of when the aldehyde group is a constituent. Here, it gets the number 1 because it is a characteristic group. The bromo and the chloro are 3 and 5 due to the alphabetical order tiebreaker.

Our next two examples in Figure 38 are of when the compound is named as an aldehyde. Since an aldehyde is a 'characteristic' functional group, and it must be at the end of the chain, these compounds don't have a locant with the aldehyde since it must be 1. Thus, the numbering of these compounds is simple because you don't have to worry about which end to start from.

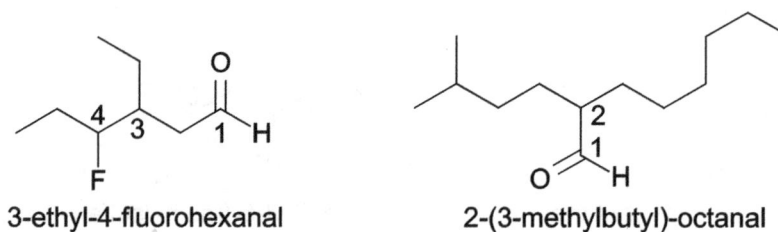

3-ethyl-4-fluorohexanal   2-(3-methylbutyl)-octanal

Figure 38: Nomenclature of compounds named as aldehydes

Ketones are similar to aldehydes except there is an -R group on either side. Ketones, like ethers and sulfides, can be named similarly in a addition to the substitutive system. Figure 39 shows a couple of simple examples with both the substitutive and common naming.

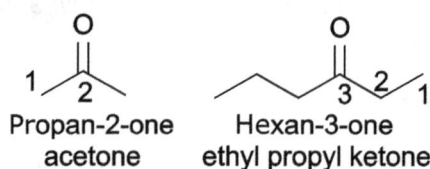

Propan-2-one   Hexan-3-one
acetone   ethyl propyl ketone

Figure 39: Nomenclature of ketones

The simplest ketone must have three carbons since one has the carbonyl group, and then each side has a methyl group. The substitutive name is 2-propanone since it is derived from propane. The common name is acetone. The other example is named hexan-3-one. Here, the locant describes the carbon where the carbonyl is attached. The second set of ketones involve a cyclic ketone and can be seen in Figure 40.

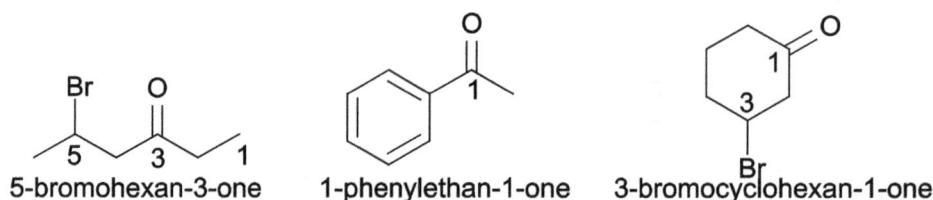

5-bromohexan-3-one   1-phenylethan-1-one   3-bromocyclohexan-1-one

Figure 40: Other examples of nomenclature of ketones

In the first example, since the compound is named as a ketone, the carbonyl is given priority in numbering, so it is carbon 3. In the second example, both the phenyl group and the ketone are associated with the same carbon, so the name is 1-phenylethan-1-one. The final example is a cyclohexanone. Again, the carbonyl carbon gets the number 1.

Carboxylic Acids, Carboxylic Acid Derivatives, and Nitriles

Carboxylic acids are a very common functional group in nature, and their reactions and those of their derivatives are important both in organic chemistry and biochemistry. In this section, there are examples of carboxylic acids, esters, amides, acid anhydrides, and nitriles. The naming of these functional groups are all slightly different. We will begin with carboxylic acids, two examples of which are shown in Figure 41. Carboxylic acids have a similar naming scheme as for aldehydes-if the carboxylic acid is part of a chain, the -e of the alkane is replaced with -anoic acid. So hexane would become hexanoic acid. If the acid is a substituent on a ring, it would become cyclohexane-1-carboxylic acid.

2,3-dimethylcyclohexane-1-carboxylic acid   4-chloro-5-fluoro-2-methylhexanoic acid

Figure 41: Nomenclature of carboxylic acids

These two examples show the naming when the carboxylic acid is denoted as a substituent and when it is part of the named molecule. The first example shows when it is a substituent, such as on a ring. The carboxylic acid will get the number 1 since it is a characteristic group. Thus the name is 2,3-dimethylcyclohexane-1-carboxylic acid. When the acid is part of the named molecule, it is numbered 1 with a similar logic to the aldehyde: It is numbered 1 because only one side has an -R group. Thus, the name for the second example is 4-chloro-5-fluoro-2-methylhexanoic acid.

Next we will tackle esters, which have a nomenclature similar to ketones and ethers. The ester is named by indicating the -R group of the alcohol side of the ester, followed by the name of the carboxylic acid modified by the suffix -oate. We see three examples in Figure 42.

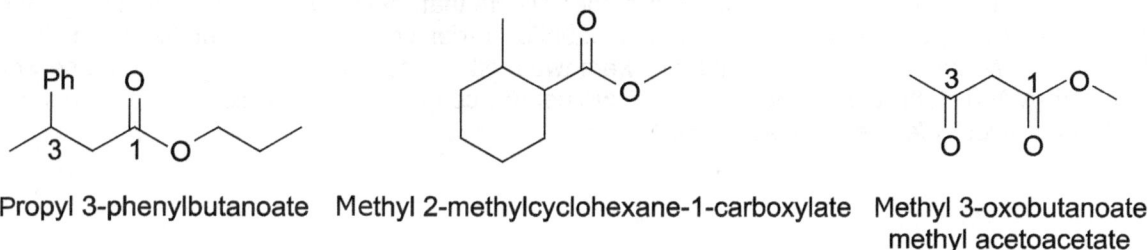

Propyl 3-phenylbutanoate   Methyl 2-methylcyclohexane-1-carboxylate   Methyl 3-oxobutanoate
                                                                       methyl acetoacetate

Figure 42: Nomenclature of esters

The first example can be named in Plain English as the propyl ester of 3-phenylbutanoic acid. The group derived from the alcohol is listed first, followed by the 3-phenylbutanoate. As with carboxylic acids, the carbonyl carbon is given the number 1. The second example is the methyl ester 2-methylcyclohexane-1-carboxylic acid. Notice how the ending here is carboxylate and not -oate. The third example is an example of a β-keto ester, used in the acetoacetic ester synthesis. The systematic name listed first is an example of when there is more than one 'characteristic group' in a molecule. This molecule has both an ester and a ketone. The ester ranks above the ketone, so the carbonyl is denoted by the prefix oxo-. The compound is the methyl ester of 3-oxobutanoic acid. As with other acids, the carbonyl carbon of the acid is noted 1, so the name is methyl 3-oxobutanoate. The common name for the compound is methyl acetoacetate.

Figure 43 has examples of an acid anhydride and an acid anhydride.
Acid halides are named similarly to esters except that the carboxylic acid is modified by adding -oyl and then putting the halide as a separate word, so 3-bromopropanoic acid chloride would become the

3-bromopropanoyl chloride   Acetic butanoic anhydride

Figure 43: Nomenclature of acid chlorides

named 3-bromopropanoyl chloride. Acid anhydrides are named similarly to ethers, where the groups on either side of the carbonyls are named in alphabetical order and then anhydride is placed after. This anhydride is derived from acetic acid and butanoic acid, so the name is acetic butanoic anhydride.

Next, we will name some amides. These examples are shown in Figure 44.

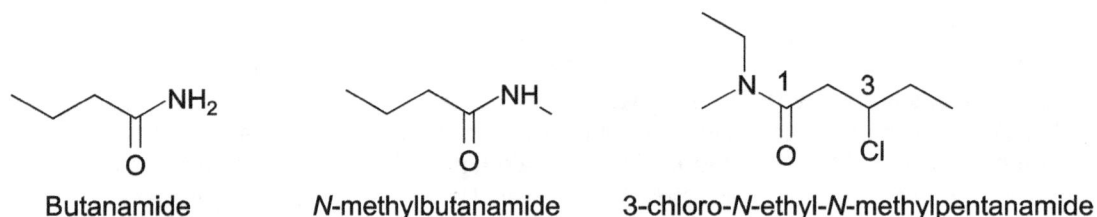

Butanamide   N-methylbutanamide   3-chloro-N-ethyl-N-methylpentanamide

Figure 44: Nomenclature of amides

These are named similarly to esters, except that the nitrogen can have up to two substituents. If there are no substituents, there is no indication in the name, and the name of the carboxylic acid is modified with -amide, so butanoic acid becomes butanamide for the first example. The second example has a methyl group on the nitrogen, and the name reflects that, being N-methylbutanamide. If there are two groups, they are named separately in alphabetical order, so the name for the last example is 3-chloro-N-ethyl-N-methylpentanamide. If there were two methyl groups on the nitrogen the name would be 3-chloro-N,N-dimethylpentamide. Finally, we will name a couple of nitriles, which are named similarly to carboxylic acids. We see examples in Figure 45.

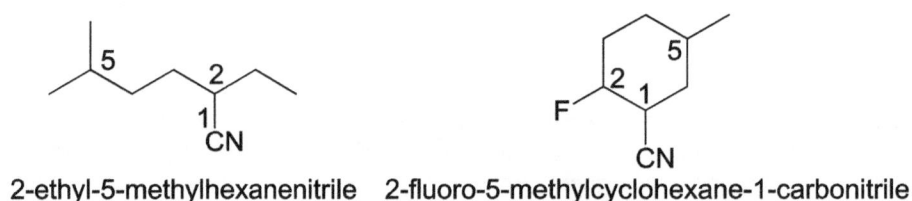

2-ethyl-5-methylhexanenitrile   2-fluoro-5-methylcyclohexane-1-carbonitrile

Figure 45: Nomenclature of nitriles

In the first example, since the nitrile is part of the chain, the carbon of the nitrile gets the number 1. Thus, for the first example, the name is 2-ethyl-5-methylhexanenitrile. In the second example, the nitrile is a substituent and also gets the number 1, so the name is 2-fluoro-5-methylcyclohexane-1-carbonitrile. Note how this is similar to the carboxylic acid nomenclature.

Amines

Our last functional group example are amines. These are compounds with a nitrogen and three H's or -R groups attached. Our first set contains a primary amine, a secondary amine, and a tertiary amine. We can see them in Figure 46. These are also characteristic groups and the -e of the alkane is replaced with -amine.

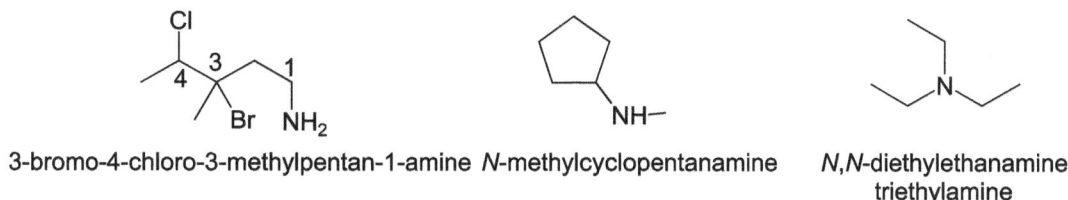

3-bromo-4-chloro-3-methylpentan-1-amine   *N*-methylcyclopentanamine   *N,N*-diethylethanamine
triethylamine

Figure 46: Nomenclature of amines

The first structure is a primary amine because the nitrogen has one -R group attached to it. Since an amine is a characteristic group, the numbering starts closest to the nitrogen. Thus, the name of the first structure is 3-bromo-4-chloro-3-methylpentan-1-amine. The second example is a secondary amine since the nitrogen is attached to two carbons. As noted previously, rings rank higher than chains so the base name is cyclopentamine. The compound is named as a substituted cyclopentanamine. The third example is a tertiary amine where there are three -R groups attached to the nitrogen. In this example, all the -R groups are the same (in this case, ethyl). The name of this compound is *N,N*-diethylethanamine. The common name for this compound is triethylamine.

The second and last set of amines are shown in Figure 47. The key takeaway from these two examples is that the rings rank higher than chains regardless of ring size. The second example is notable in that the chain is eight carbons, but the base name is still cyclobutanamine.

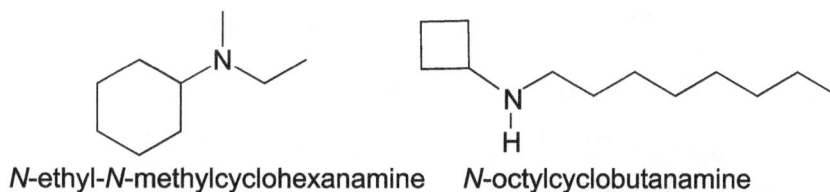

N-ethyl-*N*-methylcyclohexanamine   *N*-octylcyclobutanamine

Figure 47: Other nomenclature of amines

Nomenclature Practice

Now that we have seen a number of different examples of nomenclature and naming, it is your turn to try your hand both types of questions-giving a name for a structure, or drawing the structure of a named compound. One thing to note is that each example will have a maximum of one characteristic group within it. The format for these questions will be open response. Answers will be shown on the pages immediately following the questions in Figure 48 and Figure 49. Please be sure to attempt all the questions before looking at the answer.

Generate a systematic name for the following structures:

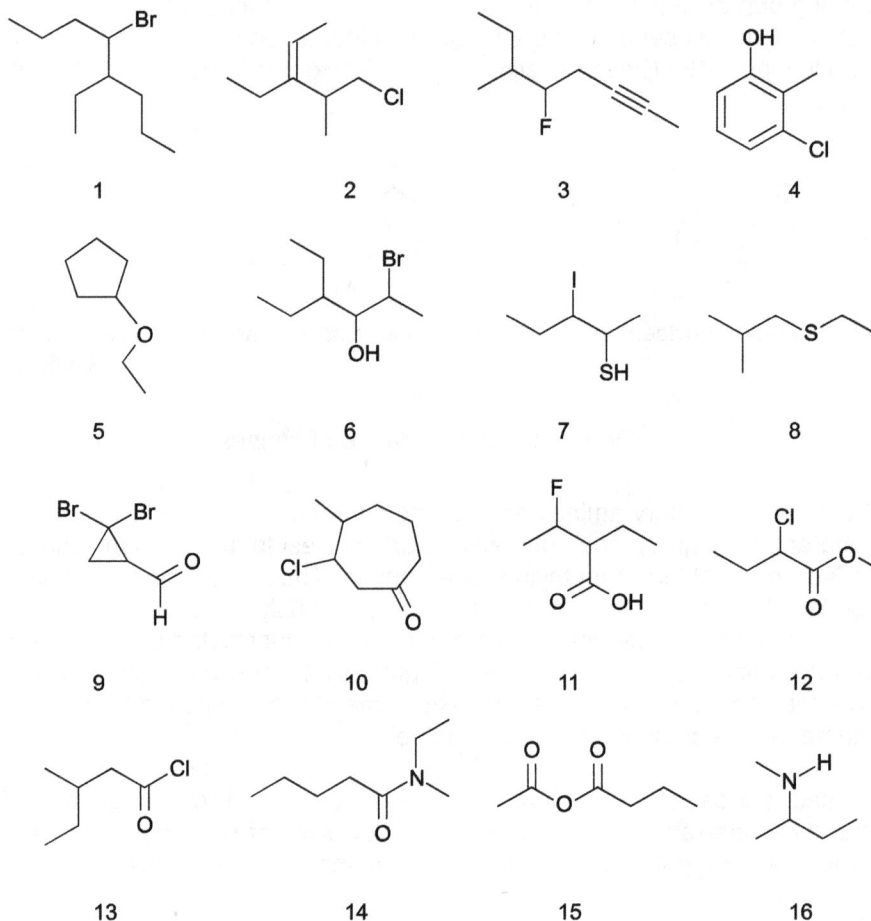

Figure 48: Nomenclature practice set

Draw a structure corresponding to the following names:

A. 1-bromo-3-chloro-2-methylpentane
B. (2E)-6-chloro-5-fluoro-3-methylhex-2-ene
C. 4-methylhept-1-yne
D. 4-bromo-3-ethylphenol
E. [(propan-2-yl)oxy]cyclopentane
F. 4-methyloctan-4-ol
G. 3-(bromomethyl)hexane-2-thiol
H. 2-methyl-3-(methylsulfanyl)pentane
I. 2-chloro-4-methylpentanal
J. 2-methylpentan-3-one
K. 4-bromo-2-methylpentanoic acid
L. methyl 2-ethyl-3-methylbutanoate
M. 5-bromo-3,4-dimethylpentanoyl chloride
N. *N*-methyl-*N*-propylpropanamide
O. 2-bromobutanoic 3-chloro-2-methylpropanoic anhydride

P. *N*-ethyl-2-methylpropan-1-amine

Answers to naming 1-16:

1. 4-bromo-5-ethyloctane
2. (2*Z*)-5-chloro-3-ethyl-4-methylpent-2-ene
3. 5-fluoro-6-methyloct-2-yne
4. 3-chloro-2-methylphenol
5. ethoxycyclopentane
6. 2-bromo-4-ethylhexan-3-ol
7. 3-iodopentane-2-thiol
8. 1-(ethylsulfanyl)-2-methylpropane
9. 2,2-dibromocyclopropane-1-carbaldehyde **10.** 3-chloro-4-methylcycloheptan-1-one
11. 2-ethyl-3-fluorobutanoic acid **12.** methyl 2-chlorobutanoate
13. 3-methylpentanoyl chloride
14. *N*-ethyl-*N*-methylpentanamide
15. acetic butanoic anhydride
16. *N*-methylbutan-2-amine

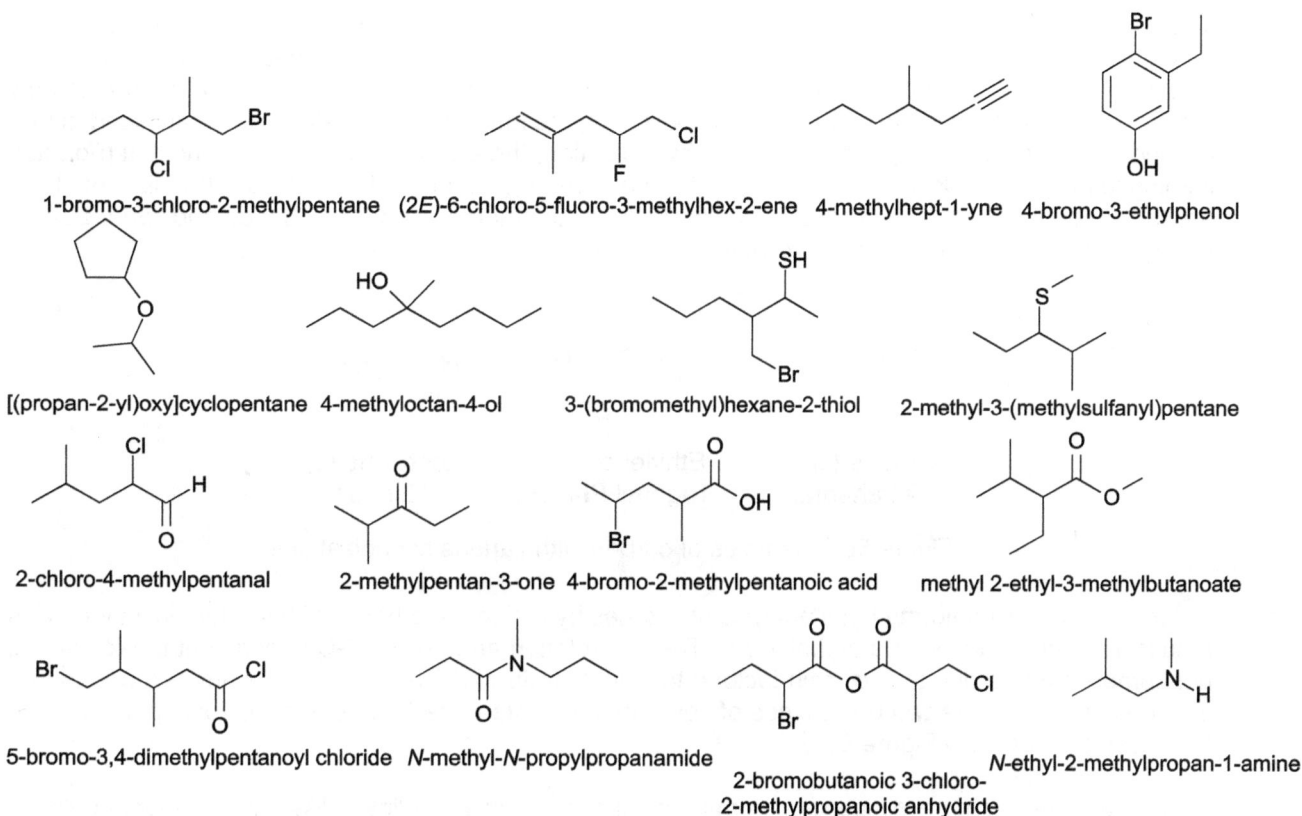

1-bromo-3-chloro-2-methylpentane (2*E*)-6-chloro-5-fluoro-3-methylhex-2-ene 4-methylhept-1-yne 4-bromo-3-ethylphenol

[(propan-2-yl)oxy]cyclopentane 4-methyloctan-4-ol 3-(bromomethyl)hexane-2-thiol 2-methyl-3-(methylsulfanyl)pentane

2-chloro-4-methylpentanal 2-methylpentan-3-one 4-bromo-2-methylpentanoic acid methyl 2-ethyl-3-methylbutanoate

5-bromo-3,4-dimethylpentanoyl chloride *N*-methyl-*N*-propylpropanamide 2-bromobutanoic 3-chloro-2-methylpropanoic anhydride *N*-ethyl-2-methylpropan-1-amine

Figure 49: Structures corresponding to names **A** -**P**

Physical Properties

In this chapter, we will take a look at some common properties related to structure with organic compounds. The structure and bonding section discusses the different types of bonding and orbitals of organic compounds, as well as Newman projections, particularly of ethane and butane. The formal charge section will review procedures for assigning, interpreting, and using formal charges. The melting points and boiling points section will discuss how intermolecular forces impact melting and boiling points.

Structure and Bonding

The carbon atom has one s and three p orbitals available for bonding in its valence shell. Thus, carbon has three typical hybridizations for bonding: sp^3, sp^2, and sp hybridization. These hybridizations have a large impact on the structure and properties of organic compounds, since they in part determine how the molecules "fit" together in bulk. In sp^3 hybridization, the four orbitals are all the same and lead to four σ-bonds and a *tetrahedral* geometry. In sp^2 hybridization, three orbitals are sp^2 and one is a p-orbital, leading to three σ-bonds, one π-bond, and a trigonal planar geometry. Finally, in sp hybdridization there are two sp orbitals and two p-orbitals, leading to linear geometries and two σ-bonds and two π-bonds. Examples of these structures are shown in Figure 50.

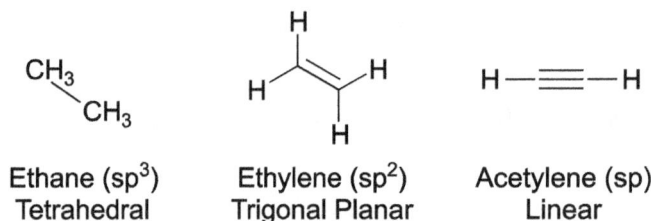

| Ethane (sp³) | Ethylene (sp²) | Acetylene (sp) |
| Tetrahedral | Trigonal Planar | Linear |

Figure 50: Examples of carbon with various hybridizations

Let's begin the conformational analysis of alkanes by looking at ethane, with sp^3 hybridization. This results in a tetrahedral shape at that atom. There is rotation around the C-C bonds. But the rotation is not completely free due to two main factors, torsional strain and steric strain. For any two bonds in a linear alkane, there are two basic types of conformations, staggered and eclipsed. Examples of these two types are shown in Figure 51.

We can see that in the eclipsed conformation, all the bonds are eclipsed like a solar or lunar eclipse. For ethane, this is the source of the rotational barrier. The torsional strain results from repulsion between the electrons in the bonds. In the staggered conformation, there is no such repulsion since the bonds do not overlap. Thus, for ethane, there are basically two states: the higher energy eclipsed and the lower energy staggered. For butane, which has four carbons, there are some additional states that exist and

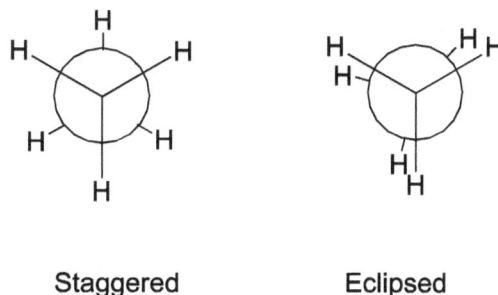

Staggered   Eclipsed

Figure 51: Staggered and eclipsed conformations of ethane

can be seen in Figure 52.

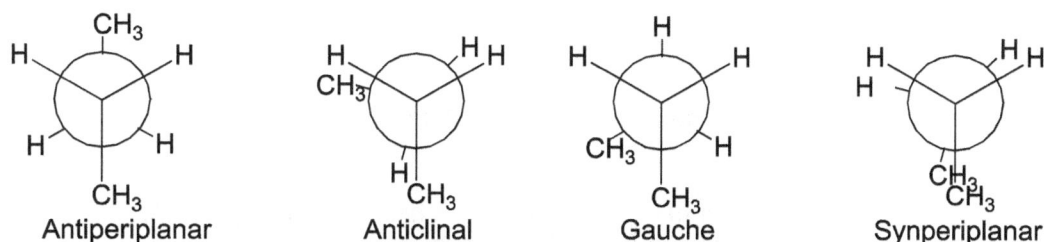

Antiperiplanar   Anticlinal   Gauche   Synperiplanar

Figure 52: Conformations of butane

For butane, there are four energetically different conformations: the antiperiplanar, the anticlinal, the gauche, and the synperiplanar. The antiperiplanar conformation is staggered with the -CH_3 groups as far apart as possible. As we rotate in 60-degree increments, the next conformation is anticlinal, which has the -CH_3 and the -H groups eclipsed. Sixty more degrees gets us to the gauche conformation. The gauche conformation is staggered, but now the methyl groups are near each other. Another 60 degrees gets to the highest energy conformation. The synperiplanar has two -CH_3 groups overlapping. Rotating around more gets us to energetically similar conformations to what we saw previously. These conformations are typical of conformations in acyclic alkanes, and the difference in energy between the antiperiplanar and the synperiplanar is around 20 kJ/mol. As a bit of context, the energy available for rotation at room temperature is around 84 kJ/mol. So at room temperature, there is free rotation for most C-C bonds. This has particular impact for NMR spectroscopy since spectra are impacted by the temperature. At room temperature, the rapid free rotation means the signals we see are an average of the conformations.

Next, let's look at sp^2 hybridization. As mentioned, the s orbital and two p orbitals are mixed, resulting in three sp^2 orbitals, which make σ-bonds and an unhybridized p orbital, which makes a double bond. In contrast to alkanes, alkenes are unable to rotate around the C-C double bond. This also has an impact on structure, and thus reactivity. The hybridization gives a trigonal planar geometry. Alkenes are a versatile reactant due to the π-bond, and a large number of other functional groups (expoxides, alcohols, carboxylic acids, alkyl halides, diols, etc.) can be generated from alkenes and carbonyl compounds, which both contain an sp^2-carbon atom.

Finally, let's briefly look at sp hybridization. One s- and one p-orbital are mixed, giving a linear molecule with two different π-bonds resulting from the two unhybridized p orbitals. The two p-orbitals react independently of one another, resulting in the ability to generate products that have either sp^2 or sp^3 hybridization. Carbon dioxide, a very important molecule, has its carbon sp hybridized. Nitriles are also a functional group that is sp hybridized and can be converted into a variety of other functional groups (including but not limited to carboxylic acids, ketones, amides, and amines).

Formal Charge

For the common elements seen in undergraduate organic chemistry (carbon, nitrogen, oxygen, and the halogens), patterns emerge with how they tend to bond when the atoms are neutral. Carbon, with four valence electrons in its outer shell, makes four bonds. Nitrogen, with five electrons, make three bonds, oxygen with six electrons, makes two bonds, and the halogens, with seven electrons, make one bond. The related concept of formal charge is also useful to review here. Formal charge is a method that gives a calculated charge but can differ from the *actual* charge. One way to think about this is to think about formal education versus actual education. If someone has a college degree, that means they met the formal requirements to graduate. However, the *real* amount of knowledge and skills gained will be determined case by case. When computing formal charges, the following procedure can be followed, assuming a correct structure has been drawn:

1. For each atom, calculate the number of valence electrons associated with that atom. This is the number of electrons that atom needs to have a zero formal charge.

2. For each atom, assign the number of electrons that atom has by assigning the atom one electron for each bond that atom has, and assign two for each lone pair.

3. Compare the number assigned to the number that the atom needs. If they are equal, there is no formal charge. If there are more electrons assigned than the atom needs, the atom will have a formal charge that is a negative number. If there are fewer electrons, then the atom will have a formal charge that is a positive. Three examples of formal charges are shown in Figure 53.

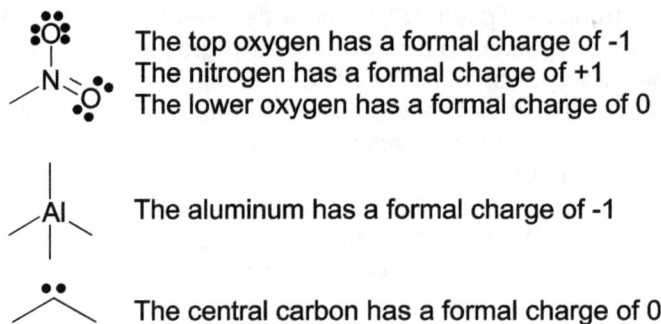

The top oxygen has a formal charge of -1
The nitrogen has a formal charge of +1
The lower oxygen has a formal charge of 0

The aluminum has a formal charge of -1

The central carbon has a formal charge of 0

Figure 53: Examples of Lewis structures and formal charges on selected atoms

For the first example, nitromethane, we will look at the charges on the nitrogen and the two oxygens. The top oxygen has six valence electrons and is assigned seven valence electrons. Six of them come from the three lone pairs and one from the N-O bond. As the number of assigned electrons is greater than the number of valence electrons, that oxygen gets a -1 charge.

In the second example, the aluminum has three valence electrons, and is assigned four electrons based on the four Al-C bonds. Thus, this also has a negative charge.

Finally, we have the structure of a carbene. Note in this example that there is no C-H bond on the central carbon. Thus, the carbon has four electrons, two from the lone pair, and two from the two C-C bonds. The central carbon has a formal charge of zero, even though there are only two bonds to that particular carbon.

One thing to remember from the formal charge versus actual charge is that the formal charge assumes that every bond is a pure covalent bond. However, in reality this is often not the case since many bonds are ionic or polar covalent. A good example of this issue is tetrafluoroborate, BF_4^-, a common weakly coordinating anion. As can be seen in Figure 54, the formal negative charge is on the boron, yet the Pauling electronegativity of F is 4.0, and that of B is 2.0.

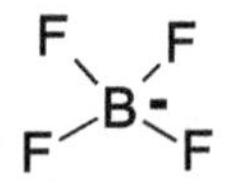

Figure 54: The tetrafluoroborate anion

The B-F bonds are all strongly polarized toward the F. The actual charge distribution is closer to the fluorines splitting the negative charge instead of the B. While this is a fairly extreme example, it suggests that most of the time the formal charge does not match the actual charge since electronegativities are often unequal for atoms in a bond.

Common questions that arise from structure and bonding topics include bond angles, molecular geometries, and formal charges.

Melting Points and Boiling Points

Other properties that are tied to the structure of a molecule are its melting point (mp) and boiling point (bp). These two points define phase changes from solid to liquid and liquid to gas. They both also depend on the contributions of four basic intermolecular forces. They are listed here in order of decreasing strength: ionic interactions, hydrogen bonding, dipole-dipole interactions, and London forces (Van der Waals forces). We will discuss each of them in order. Again, it should be noted that the mix of these will vary by the individual compound.

1. **Ionic interactions.** For organic compounds, ionic interactions are typically going to be between alkali or alkaline earth metal cations and an organic anion. The ionic bond is very strong, and so these salts will tend to have significantly higher mp and bp than their neutral counterparts. For example, sodium acetate has a mp of 324°C and a bp of around 881°C compared to a mp of 17°C and a bp of 118°C for acetic acid. While ionic interactions are quite strong, most organic compounds that are isolated are neutral, so they other three forces tend to play a larger role in most cases.

2. **Hydrogen bonding.** These interactions occur in compounds that contain H-F, O-H, or N-H bonds. The electronegativity difference between H and N, O, and F is what makes this happen. The Pauling electronegativity of H is 2.2, while N is 3.0, O is 3.4, and F is 4.0. Even though Cl has a similar electronegativity to N, it is not considered to form hydrogen bonds since it is a third-row element and it is larger, physically preventing a close association with H, as well as more polarizable versus fluorine. When present, hydrogen bonding can have significant impacts on these properties. A prime example is the difference between the two constitutional isomers ethanol and dimethyl ether. Both have the formula C_2H_6O. Ethanol has a mp of -114°C and a bp of 78°C, while dimethyl ether has a mp of -141°C and a bp of -24°C. Since these are consitutional isomers, the differences in mp and bp are due to structural differences, and the biggest difference is the fact that ethanol can hydrogen bond, and dimethyl ether

cannot.

3. **Dipole-dipole interactions.** These interactions are similar to hydrogen bonds except they are weaker and deal with bonds of carbon with more electronegative atoms instead of hydrogen. Since carbon is larger and less electronegative than hydrogen is (EN 2.5 for C versus EN of 2.2 for H), these interactions are weaker. An example is the difference between the two constitutional isomers butan-2-one and ethyl vinyl ether. Both have the formula C_4H_8O. Butan-2-one has a mp of -86°C and a bp of 80°C, while dimethyl ether has a mp of -116°C and a bp of 33°C. Since these are constitutional isomers, the differences in mp and bp are due to structural differences, and the biggest difference here is the fact that the ketone is a more polar functional group than the alkene.

4. **London dispersion forces.** These are the weakest of the interactions. These occur when the electron density around molecules spontaneously becomes unbalanced. That is, on average, the valence electrons are equally distributed around an atom. However, at any moment, there may be a momentary imbalance, generating a temporary dipole. These temporary dipoles are the dominant intermolecular force for hydrocarbons and other compounds with little or no dipole. The more atoms in a molecule, the more London forces there are due to the larger surface area. Thus, for linear hydrocarbons, the mp and bp generally increase with increasing molecular weight. We can see data for the first 10 hydrocarbons in Table 3:

Table 3: bp and mp for the First 10 Linear Alkane Isomers

Hydrocarbon	mp°C	bp°C
methane	-182	-164
ethane	-183	-89
propane	-190	-42
butane	-138	-1
pentane	-130	36
hexane	-95	69
heptane	-91	98
octane	-57	125
nonane	-54	150
decane	-30	174

As we look back at our examples, the bp are more easily explained by these forces than the mp, and that is because the mp is also impacted by how well the molecules can pack together. In other words, going from a liquid to a gas is qualitatively different from going from solid to liquid. One reason is that the density generally decreases substantially when going from liquid to gas, and the same is *not* seen when going from solid to liquid. As we can also see in the table, the mp do trend upward with number of carbons, but not as smoothly as for the bp. For instance, propane has the lowest mp of the alkanes listed. Also, there is a pattern of the melting points going up slightly when going from even to odd, and going up a larger amount when going from odd to even. This is indicative of the better packing the even-numbered alkanes have versus the odd-numbered counterparts. Another trend we see is that as surface area decreases (generally from branching) the mp and bp decrease for a series of hydrocarbon constitutional isomers. Table 4 has data for some isomers of heptane.

There are a few things we can take from the information we have seen about intermolecular forces:

1. **The trends in mp and bp do not vary in the same way.** Going from solid to liquid and liquid to gas

Table 4: bp and mp of Selected Isomers of Heptane

Hydrocarbon	mp°C	bp°C
heptane	-91	98
2-methylhexane	-118	90
3-methylhexane	-119	92
2,2-dimethylpentane	-124	79
2,4-dimethylpentane	-119	80
3,3-dimethylpentane	-135	86

are different processes.

2. **The molecular weight of a molecule doesn't have as big of an impact as we might think.** The ionic and hydrogen-bond intermolecular forces have a larger impact on these properties than dipole-dipole or London forces.

3. **The way the molecules pack together can have a large impact on these properties.** For dipole-dipole and london forces, the surface area matters significantly.

Questions that involve physical properties usually have you consider a series of compounds that are different in one of three ways: (1) intermolecular forces, (2) MW, or (3) surface area.

Mass Spectrometry and Infrared Spectroscopy

Imagine you are in lab, and you have isolated a beautiful yellow crystalline solid. This is an unknown product to you, and so the question arises "How do I know what this compound is?" This chapter and the next deal with how we can identify what these compounds are using various instruments. This chapter deals with using mass spectrometry (MS) and infrared spectroscopy (IR). We will see some examples of compounds with particular functional groups to illustrate the basics of using MS and IR.

Mass spectrometry is a technique that can be used to formulate at least one very important piece of information for an unknown compound: the molecular formula. The molecular formula can then be used to calculate a number of unsaturations, which when used with the other common techniques (infrared and nuclear magnetic resonance spectroscopy) allow for elucidation of the structure. MS is a destructive technique– that is the sample used to generate the spectrum is destroyed, and thus it is non-recoverable. We are going to look at the output of the instrument, and not so much the particulars of the instrument. Very briefly, the sample is introduced into the instrument, where the bonds are broken apart using various ways to transfer energy and then the fragments that are positively charged are sorted by their mass/charge (m/z) ratio. Sophisticated instruments are capable of giving very accurate mass, which can be used to generate molecular formulas.

Besides the formula, some practical aspects of mass spec are looking for isotopic signatures for Br, Cl, S, and I. This text will not get into the details of the fragmentation patterns. There are several notable aspects of a spectrum. First is the base peak, which is simply the tallest peak in the spectrum. Then there is the M+ peak, which is the peak corresponding to the m/z of the compound minus an electron (to get the positive charge). For Br, which has two main isotopes (^{79}Br and ^{81}Br) that exist in roughly equal amounts, there will be two peaks at M+ and M+2, which are of almost equal height. We can see an example of this for the spectrum of 1-bromopropane in Figure 55. We can see the two signals of almost identical height indicating bromine at m/z 122 and 124.

A similar situation exists for Cl, which also has two isotopes (^{35}Cl and ^{37}Cl) that are present in roughly a 3:1 ratio. We can see an example of this in the MS of 2-chloro-2-methylpropane in Figure 56. Note the peaks at m/z 77 and m/z 79 that show the 3:1 ratio.

Sulfur is similar with ^{34}S and ^{32}S. For sulfur, the key pattern is an M+ and M+2 peak that are in a 25:1 ratio, which is noticeable but small. We can see an example of this in the spectrum of dimethylsulfide in Figure 57. Notice the two peaks at m/z 62 and 64.

Finally, iodine tends to fragment an generate a positively charged I at m/z 127. This is often a small peak, but still noticeable. We can see an example of this for iodomethane in Figure 58.

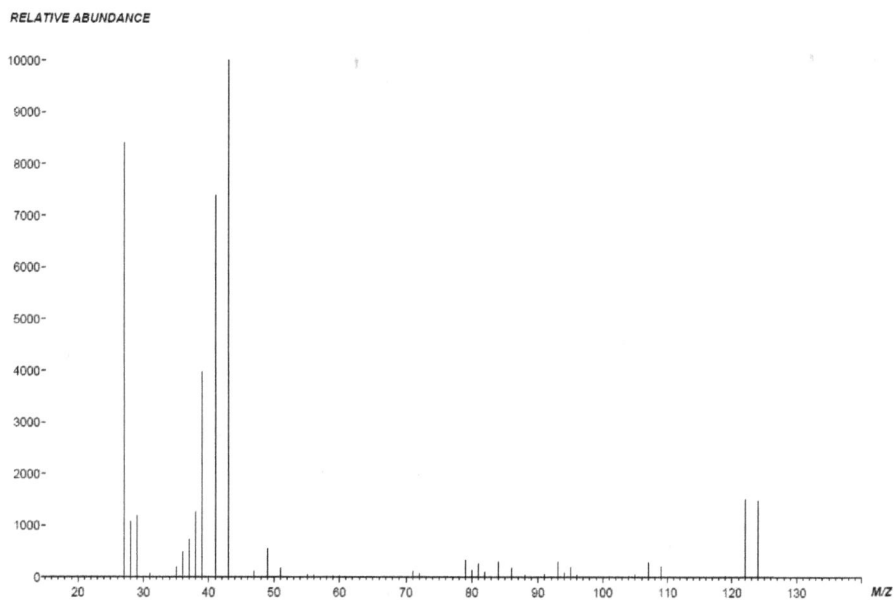

Figure 55: Mass spectrum of 1-bromopropane

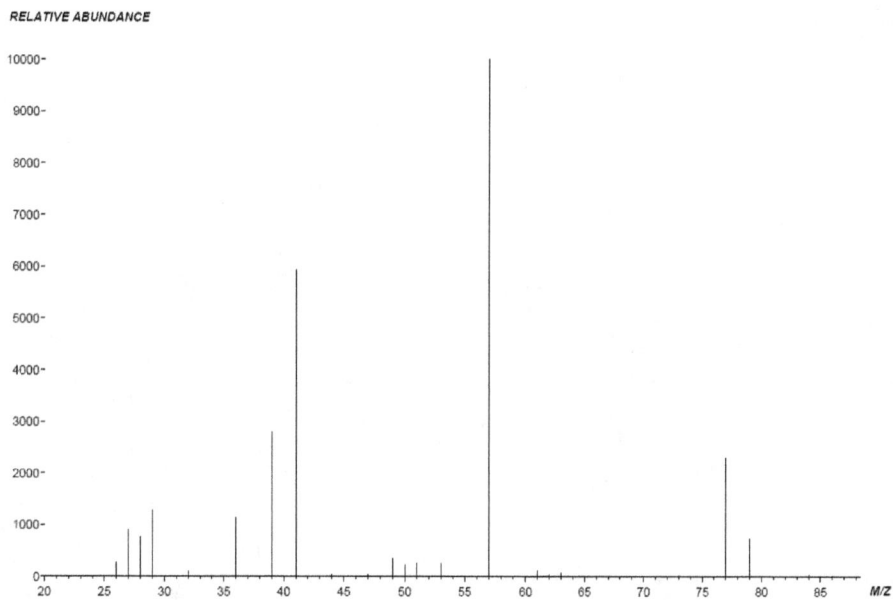

Figure 56: Mass spectrum of 2-chloro-2-methylpropane

RELATIVE ABUNDANCE

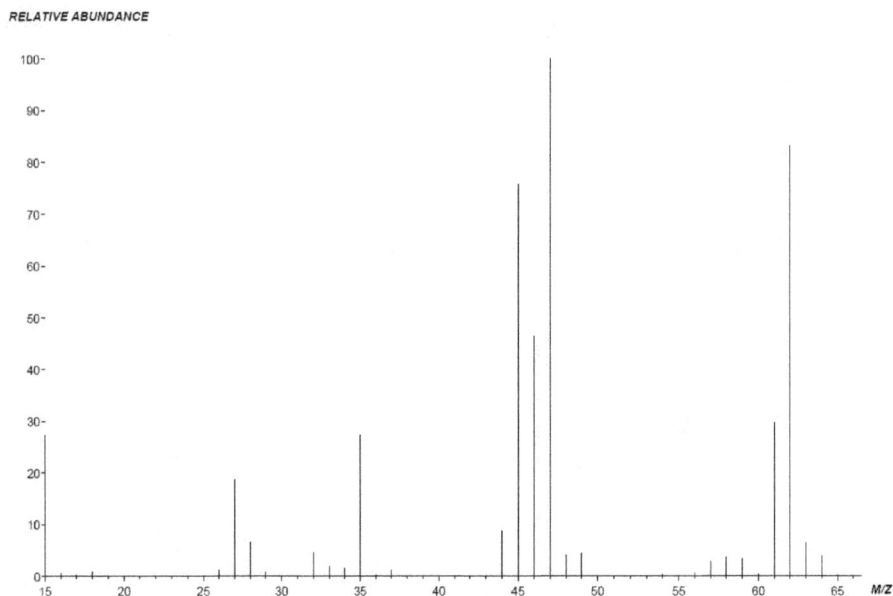

Figure 57: Mass spectrum of dimethylsulfide

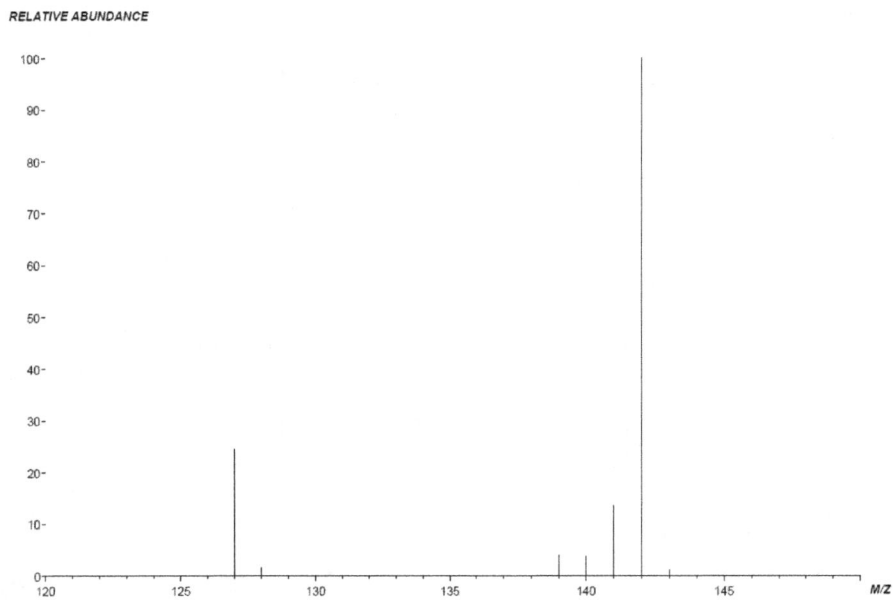

RELATIVE ABUNDANCE

Figure 58: Mass spectrum of iodomethane

There are many other uses for MS, particularly for large molecules since they generate MS that are like fingerprints, but for this book, we will focus on NMR and IR.

Infrared spectroscopy is still used as a characterization technique because taking an IR of most organic compounds only takes a minute or so depending on the sample. Also, in many cases the sample can be recovered. As the name indicates, IR spectra use frequencies in the IR region of the electromagnetic spectrum. Spectra are typically arranged with wavenumbers (cm^{-1}) on the X-axis, and transmission on the y-axis. Thus, if a sample strongly absorbs at a particular wavenumber, the peak will point *downward*, which is the opposite of what we see for MS and NMR spectra. The units on the X-axis are proportional to the frequency, which means that the larger the number, the more energy needed for bond vibration to occur.

Thus, the IR spectrum can be split up into a few main regions:

4000 to 3500 cm^{-1}: This region is where O-H and N-H bonds typically absorb.

3200 to 2800 cm^{-1}: C-H bonds (present in almost all organic compounds) show up in this region. Depending on the precise functional group, these peaks may or may not be diagnostic.

2300 to 2100 cm^{-1}: This region contains bonds that are sp-hybridized, including the functional groups of alkynes, nitriles and carbon dioxide among others.

1800 to 1500 cm^{-1}: This are contains absorbtions for a number of double bonds, including C=N, C=O, and C=C.

Once we get below this region, the shifts become less diagnostic for organic compounds, and below 1000 cm^{-1} is generally called the fingerprint region since there are a number of overlapping signals that are similar to fingerprints. We will now look at a number of IR spectra and look at some of the particular functional groups in each structure. There are often questions involving unknown compounds where you are given a formula, and then IR and NMR information and asked to find or generate a structure. Knowing the number of unsaturations can give the IR data context and help indicate potential functional groups in the molecule.

First, we will look at an example of an alkyl bromide to begin. This can be seen in Figure 59.

Csp^3-H bonds' characteristic signal is a stretch around 3000 cm^{-1}. This group of signals is very common in organic molecules and won't be pointed out in the rest of the spectra. Otherwise, there is nothing remarkable about this spectrum, as the C-Br bond signals are below 1500 cm^{-1}. Next, let's look at an IR spectrum of an alkene, which is shown in Figure 60.

Here we see the Csp^2-H stretch at 3038 cm^{-1} and the C=C stretch at 1649 cm^{-1}.

Next, let's look at a primary amine, 2-aminopropane. This spectrum can be seen in Figure 61.

Here, the interesting feature is the forked appearance of the primary amine group. One signal is at 3358 cm^{-1}, the other at 3277 cm^{-1}. IR spectra of primary amines have one spike instead of two in the same region, and spectra of tertiary amines have no signals in this region. Next, let's look at the IR spectrum of a carbonyl compound, in this case an aldehyde. This can be seen in Figure 62.

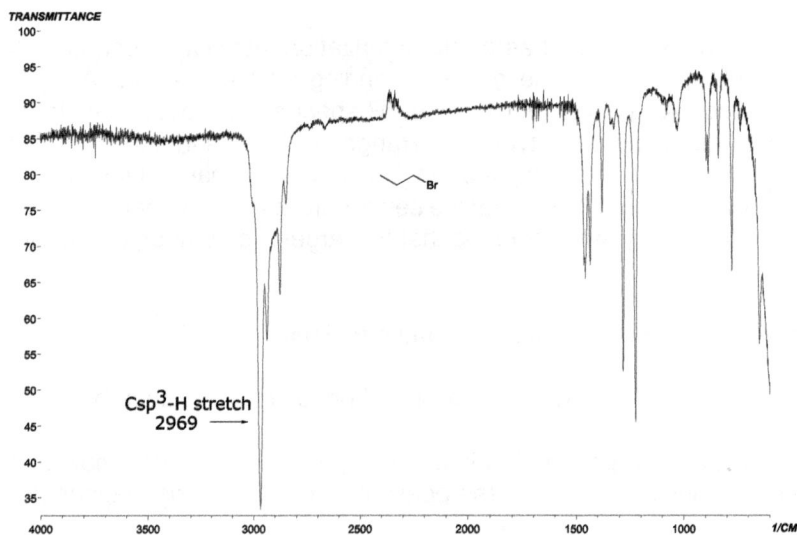

Figure 59: IR spectrum of 1-bromopropane

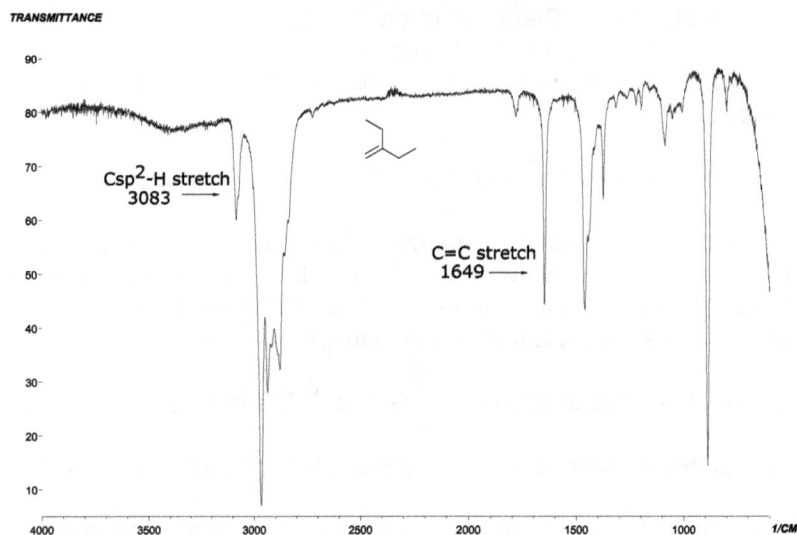

Figure 60: IR spectrum of 2-ethylbut-1-ene

TRANSMITTANCE

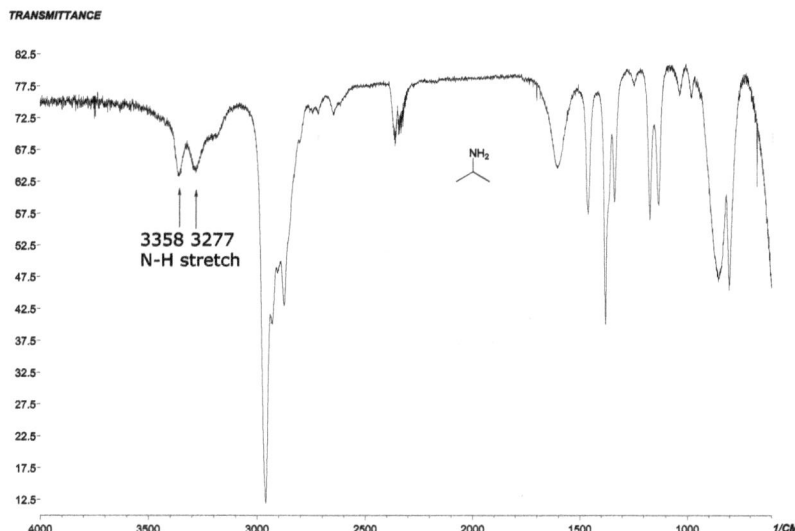

Figure 61: IR spectrum of 2-aminopropane

Particularly noteworthy in the IR spectrum of aldehydes are the two peaks at 2830 and 2750 cm^{-1}. These are only seen in compounds that have a C-H attached to a carbonyl, so they are seen in aldehydes and not much else. The typical alkyl C-H stretches occur from around 3300 cm^{-1} for C-H stretches in alkynes, 3100 cm^{-1} for C-H stretches for alkenes and aromatic compounds, and around 2900 cm^{-1} for alkyl C-H stretches, so the aldehyde C-H stretches are just a bit lower than alkyl C-H stretches. Carbonyl stretches are large and prominent, and absorb around 1700 cm^{-1}. The precise location depends on what is attached to the carbonyl. In this case, the aldehyde is attached to a phenyl ring, so it is a bit less than 1700 cm^{-1}. Other examples we will see are a bit above 1700 cm^{-1}. Next, let's look at another carbonyl compound, an ester. This spectrum can be seen in Figure 63.

Here, the main features are the typical alkyl C-H stretch, and the carbonyl stretch which is higher than the aldehyde seen previously. Now, let's look at at the IR of a carboxylic acid in Figure 64.

Here, the key visual features of the spectrum are the carbonyl stretch at 1703 cm^{-1} and the very broad -OH stretch that overlaps with the alkyl C-H group to give a jagged, craggy appearance between 3500 and 2500 cm^{-1}. This is a very distinctive feature that is different from the N-H stretches of an amine (which are much more narrow) and of an alcohol, which is much smoother. Now, let's look at the IR spectrum of an alcohol below in Figure 65.

The IR of an alcohol also has a distinct feature, which is the broad, smooth peak between 3500 and 3000 cm^{-1}, which does not overlap with many C-H stretches. This does not mean that it never overlaps, just that it is not jagged like the carboxylic acid spectrum.

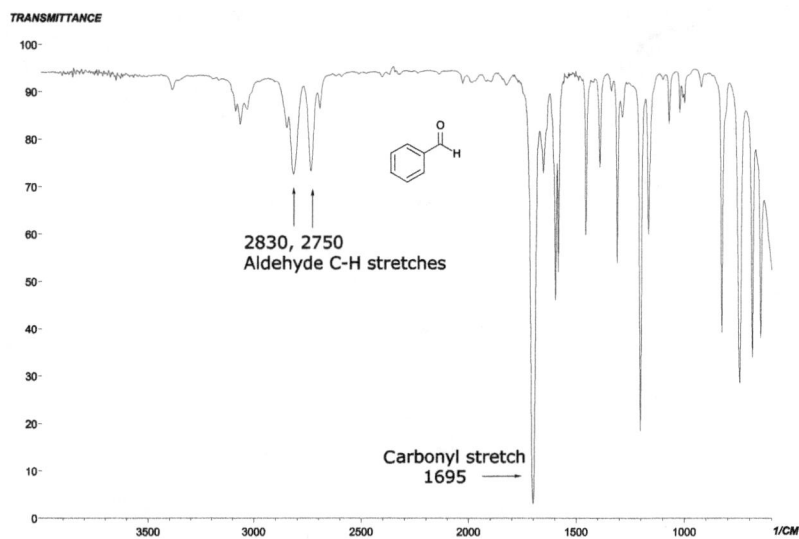

Figure 62: IR spectrum of benzaldehyde

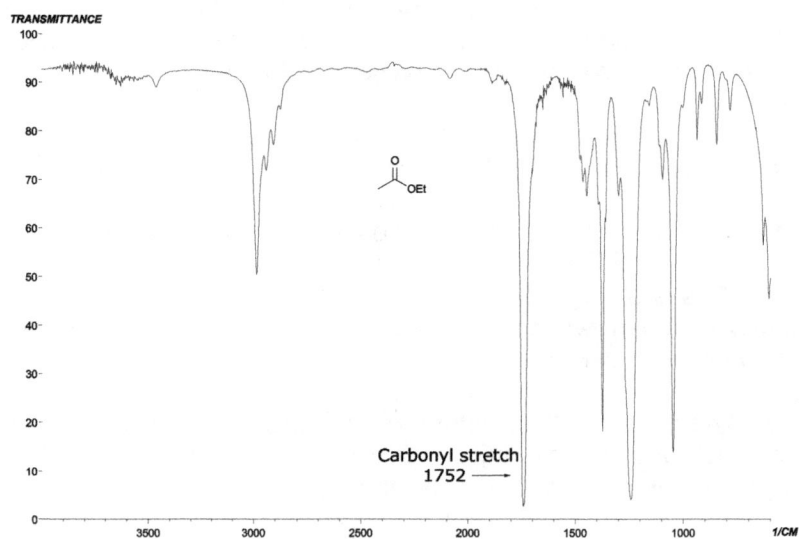

Figure 63: IR spectrum of ethyl ethanoate

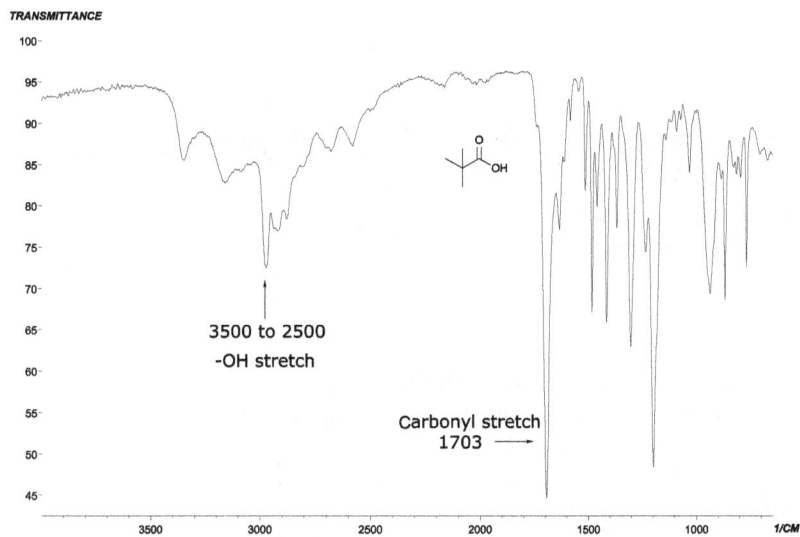

Figure 64: IR spectrum of 2,2-dimethylpropanoic acid

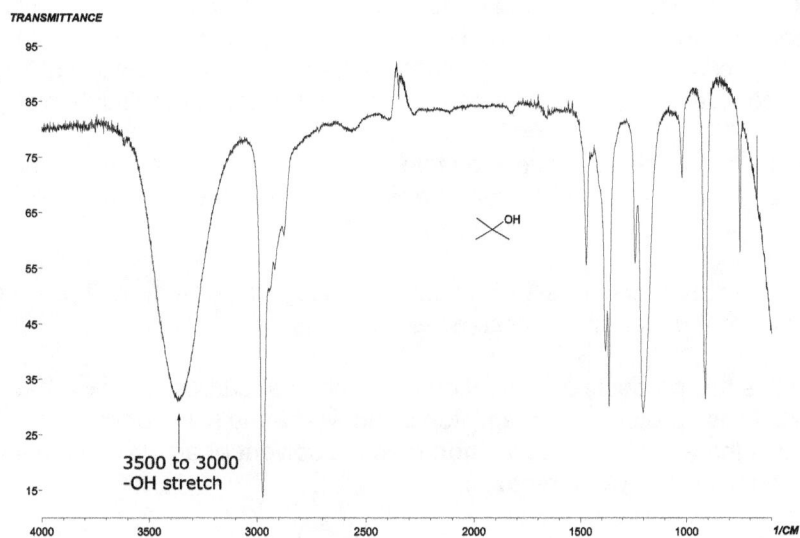

Figure 65: IR spectrum of 2-methylpropan-2-ol

Nuclear Magnetic Resonance Spectroscopy

NMR spectroscopy is essentially the same technology as the more well-known molecular resonance imaging (MRI). This technique is also nondestructive, meaning you can recover your sample. We will very briefly discuss the two most common nuclei used in NMR spectroscopy of organic compounds: ^{1}H and ^{13}C. Many other nuclei are used, but these are most common at the undergraduate level. We will look at ^{1}H NMR for the following two reasons:

1. ^{1}H NMR takes many fewer scans compared to ^{13}C using the same sample because the ^{1}H nucleus is around 99 percent of the naturally occuring hydrogen, while ^{13}C is present at around 1 percent, with the 99 percent as the NMR inactive nucleus ^{12}C.

2. Most ^{13}C NMR spectra taken are C-H decoupled. Practically, this means most of the ^{13}C signals are singlets, removing much information. This isn't to say that ^{13}C NMR is not useful, but the majority of routine NMR spectra are ^{1}H NMR.

Thus, ^{1}H NMR is a sensitive, quick, and nondestructive technique that can provide lots of information about a compound in just a few minutes. Before we talk about what information we can get about protons, we need to discuss the four different relationships protons can have: homotopic, enantiotopic, diastereotopic, and completely different. We can see examples of these relationships in Figure 66.

For all of these cases, we probe the relationship between any two protons by imagining we replace each of them with a different atom. We then compare the two products and look at the relationship between them.

(a) **Homotopic.** The two products are the exact same compound, so we term those protons homotopic, and they will be equivalent under all circumstances.

(b) **Enantiotopic.** The two products are enantiomers, since the carbon that now has the -X group has four different atoms. These protons are enantiotopic and will be the same under achiral conditions. This is mainly the case is if there is either a chiral, non-racemic solvent or additive in the solution. If there is, the protons in (b) would then be diastereotopic.

(c) **Diastereotopic.** The two products are diastereomers and will be different under all NMR conditions.

(d) **Completely different.** The two products are constitutional isomers. These will also be different under all circumstances.

An ^{1}H NMR spectrum gives three key pieces of information about protons:

(a) Replace H_a or H_b with new atom (X) — Same

(b) Replace H_a or H_b with new atom (X) — Enantiomers

(c) Replace H_a or H_b with new atom (X) — Diastereomers

(d) Replace H_a or H_b with new atom (X) — Different compounds

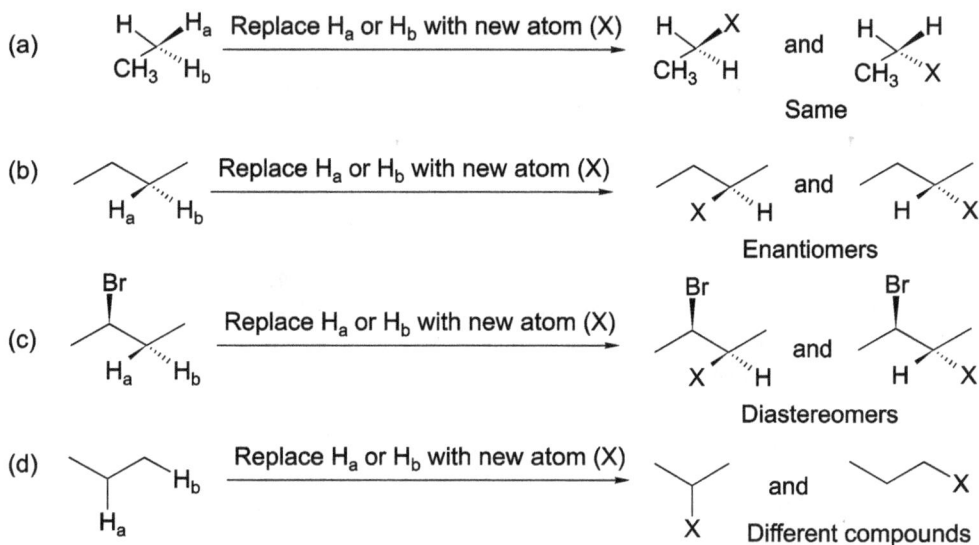

Figure 66: Relationships of various pairs of protons

1. Chemical shift. This is the measure of how much the magnetic field is interacting with the proton. The larger the chemical shift, the more deshielded. See Table 5 for examples of typical chemical shifts for proton types.

2. Integration. This simply tells us how many protons are in a particular environment. For example, a methyl group would integrate for three protons. Keep in mind, these are often reported as ratios, with the smallest integration set to one. If you have the formula, this may point to a multiple of the integrations.

3. Splitting. These patterns tell us how many protons are on a neighboring atom, usually carbon. The n+1 rule generally applies here. For example, if there are no protons on a neighboring atom, then the signal would be a singlet (0+1) denoted as s. Other typical splitting patterns are shown in Table 6. In addition, if there are different types of protons on multiple neighbors, then the splitting patterns become much more complex, and depending on who is looking and why they are looking at the spectra, it may be labled a multiplet, m. To identify many of the compounds on undergraduate organic exams, multiplet will suffice since they are often done by visual inspection. However, your instructor is the ultimate authority about what would constitute a multiplet. If you want more detail or want to look deeper into structural features, determining the separate splittings are worth the extra effort.

There are a number of known splitting pattern/integration combos that are diagnostic for partial structures. We can call these isolated spin systems; that is, these isolated systems are not connected to other atoms with protons. To be very specific, the terms methyl, ethyl, isopropyl, and *tert*-butyl refer to spin-systems and not the alkyl groups themselves. The chemical shift of these systems will depend on what other atoms are in the molecule. Table 6 shows a number of these groups along with their splitting and integration values. Examples and data are shown for each of these spin systems.

1. Methyl. This shows up as a singlet for 3H (s, 3H). In Figure 67, we can see the ^1H NMR spectrum for 2-methylnaphthalene. This figure does not have an inset or zoom since the signal is a singlet and there is no splitting. The other seven protons are in the aromatic region. Two other things are of note with this spectrum. Since it is a real spectrum, there are a few additional peaks: There is a small peak around 7.24 ppm that corresponds to $CHCl_3$. There is also a small peak around 1.5 ppm that corresponds to

Table 5: ^{1}H NMR Chemical Shifts for Selected Functional Groups

Proton type	Chemical shift in ppm (δ)
alkane	0.9 - 1.8
allylic	1.6 - 1.9
alpha carbon to carbonyl	2 - 2.6
benzylic	2.2 - 2.5
alkyne C-H	2.5 - 3
alkyl halide	3 - 4.5
alpha carbon to ether or alcohol	3.3 - 4
alkene C-H	4 - 6
aromatic	6.5 - 8.5
aldehyde	9 - 10.5
amine - NH_2	1 - 5 (exchangeable with D_2O)
alcohol -OH	1 - 7 (exchangeable with D_2O)
carboxylic acid -OH	10 - 12 (exchangeable with D_2O)

[1] Note that -OH and -NH protons can exchange with D_2O and disappear from the spectrum. [2] These are ranges and individual shifts may fall outside of these ranges.

Table 6: ^{1}H NMR Splitting and Integration for Selected Spin Systems

Spin system	Splitting and integration
methyl	singlet, 3H
ethyl	quartet, 2H and triplet, 3H
isopropyl	septet, 1H and doublet, 6H
tert-butyl	singlet, 9H

Note that the chemical shift of these groups will vary depending on what the spin system is attached to (i.e., shifts will be more deshielded if attached to an oxygen vs. a carbon.)

H_2O in chloroform. The rather large peak at 0 ppm is from an internal standard, tetramethylsilane.

2. **Ethyl.** The ethyl group is made of two carbons, a carbon with two protons bonded to a carbon with three protons (methyl group). The two signals we are looking for (and both need to be present) are a quartet for 2H (q, 2H) and a triplet for 3H (t, 3H). Details about this group are combined with those for the *tert*-butyl group.

3. **Isopropyl.** This group has three carbons and two signals because of the symmetry in the group. Assuming we have proton equivalency, the two signals we are looking for here are a septet for 1H (septet, 1H) because the central carbon of that group has two methyl neighbors that are symmetrical, so n+1 for those is 7. The other signal is a doublet for 6H (d, 6H) since each equivalent methyl group sees the same carbon with one proton as a neighbor. Our example (2-isopropylmalic acid) is illustrative of what happens when we have diastereotopic (inequivalent under all circumstances) protons. We can see the spectrum in D_2O in Figure 68. Notice how the two methyl groups and the protons of the -CH_2 group are not equivalent.

4. ***tert*-Butyl.** This group has four carbons, nine protons, and only one signal since it is also highly symmetrical. The *tert*-butyl group has three methyl groups attached to a single carbon, so it is a singlet

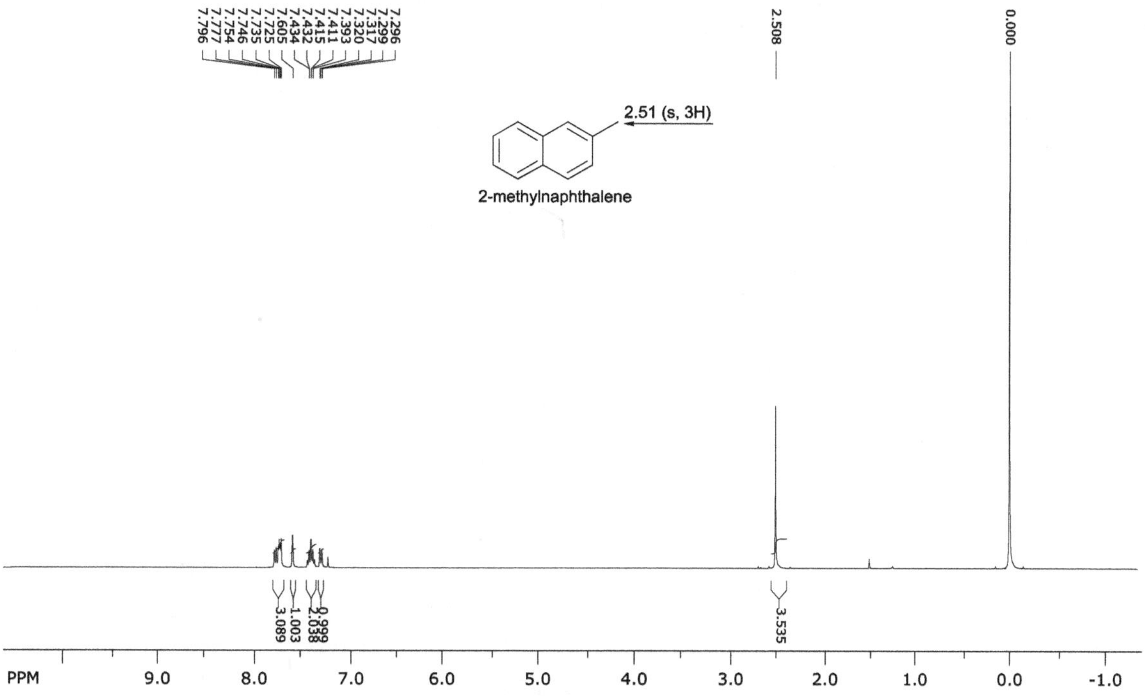
2.51 (s, 3H)
2-methylnaphthalene
PPM

Figure 67: ^{1}H NMR of 2-methylnaphthalene

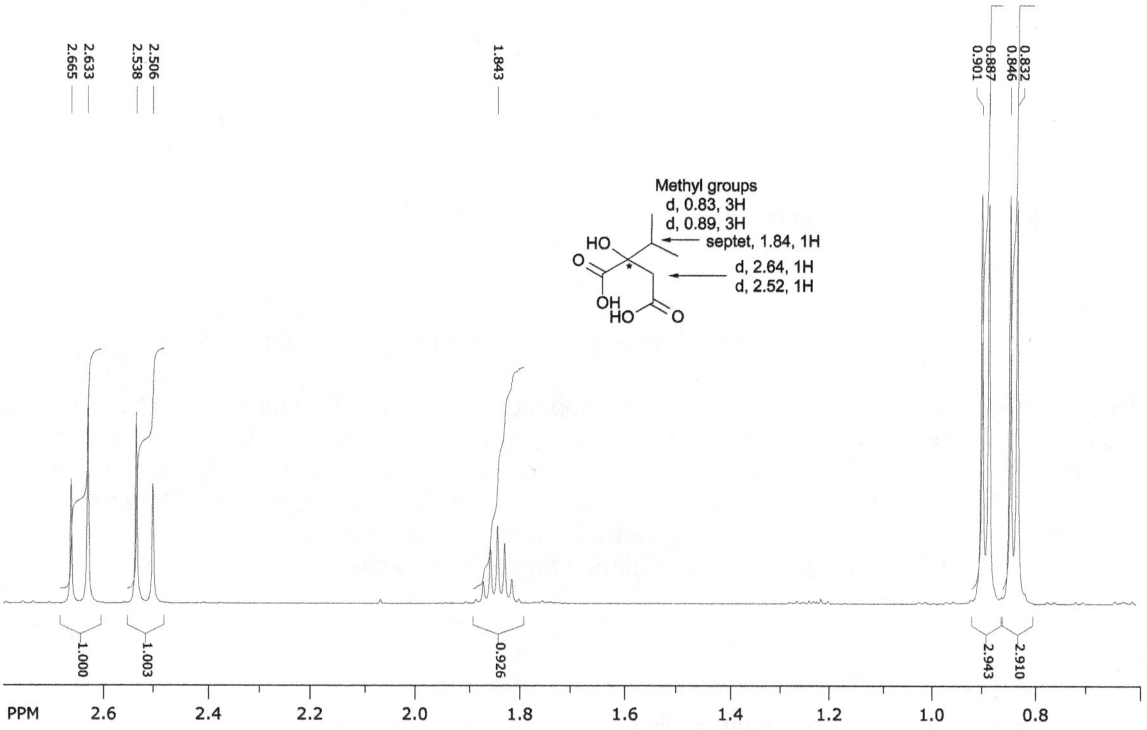
Methyl groups
d, 0.83, 3H
d, 0.89, 3H
septet, 1.84, 1H
d, 2.64, 1H
d, 2.52, 1H
PPM

Figure 68: ^{1}H NMR of 2-isopropylmalic acid

for 9H (s, 9H). Figure 69 has the full ^{1}H NMR spectrum for the compound 1-(4-Ethoxyphenyl)-3-(2-methyl-2-propanyl)urea, which contains both an ethyl group and a *tert*-butyl group.

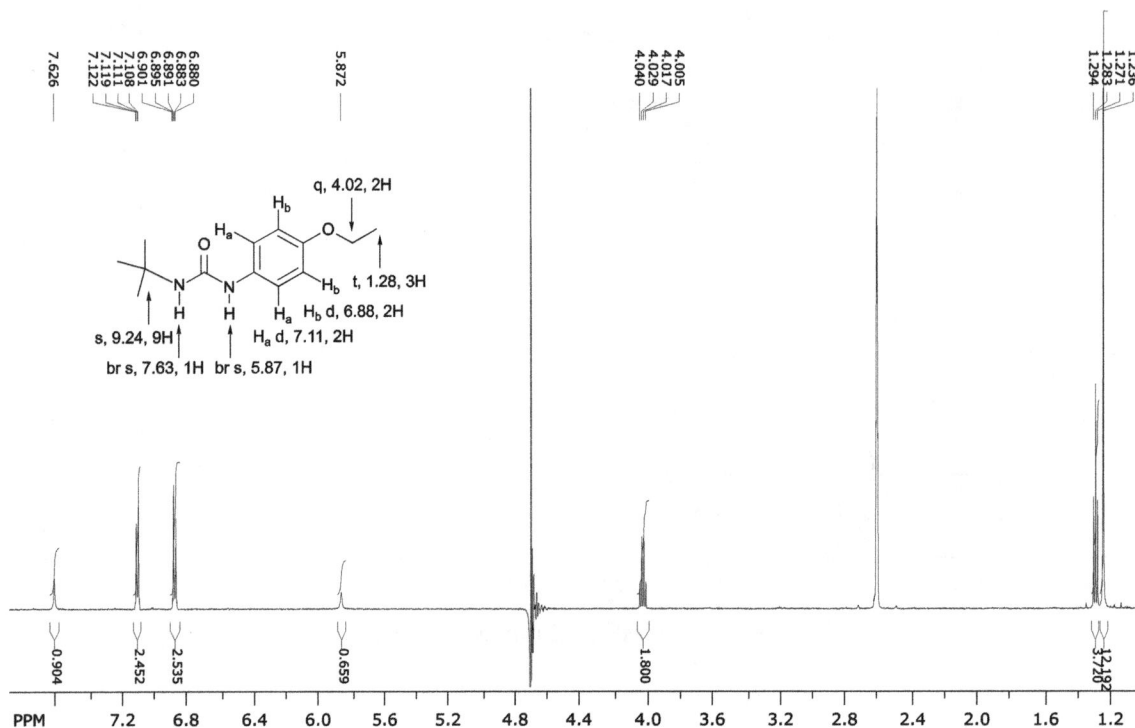

Figure 69: ^{1}H NMR of *tert*-butylethoxyphenylurea

A second figure, Figure 70 , zoomed in from 4 to 1.2 ppm, shows the ethyl group and the tert-butyl group.

Finally, let's look at an example of a typical ^{1}H NMR spectrum (from the OSDB at UNF) of 1-bromopropane in Figure 71.

If we tabulated the data for this spectrum, we would say there are three signals: 3.39 (s, 2H), 1.88 (m, 2H), and 1.00 (t, 3H). The next batch of examples will only use the tabulated spectra.

Now, let's go through some examples of compounds. The data for chemical shift for the compounds in this section come from the SDBS, a repository for spectral data for organic compounds: https://sdbs.db.aist.go.jp/sdbs/cgi-bin/direct_frame_top.cgi We will start with examples that use drawings and the tabulated data in the format of chemical shift, (multiplicity, integration). Depending on the level of detail, there may also be a coupling value in Hertz. In our case, we will generally not have the coupling values, although they can be useful, depending on the context.

Alkanes

Alkanes are a part of many organic molecules, and these make up a large part of many organic compounds. As mentioned, alkane groups make up part of many organic compounds and will not have a

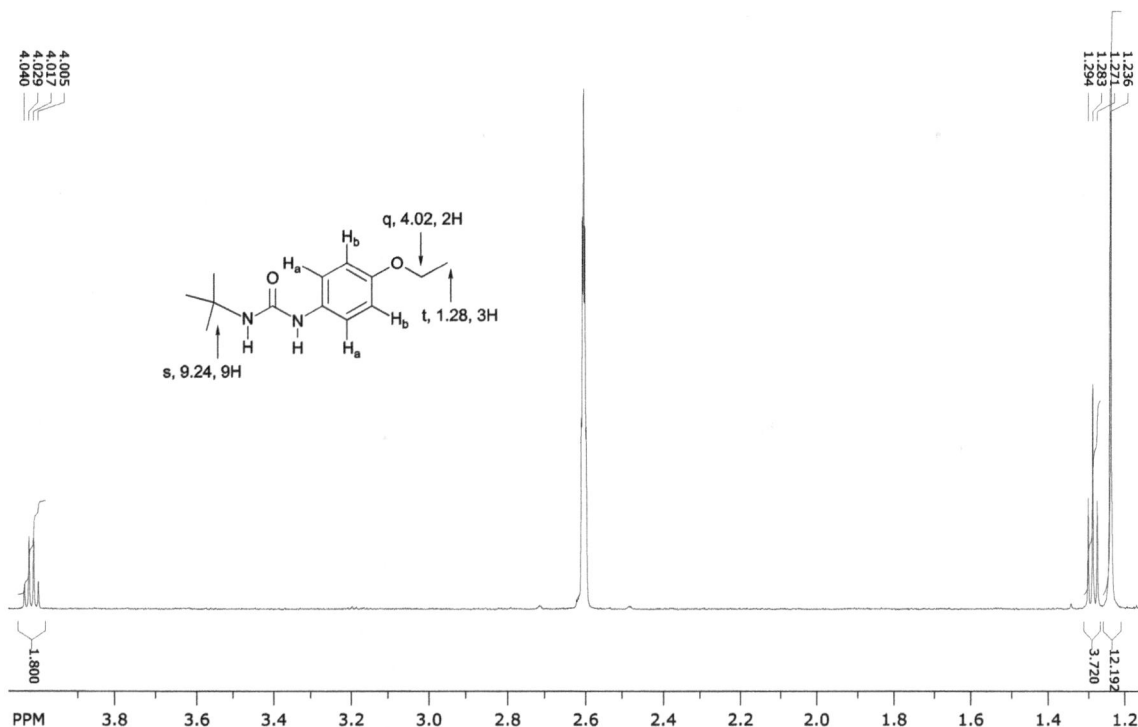

Figure 70: A zoomed-in portion of the ^{1}H NMR of *tert*-butylethoxyphenylurea

separate section. The shift range for alkanes is approximately 0.9 to 1.8 ppm, which is at the shielded end of the spectrum.

Alkenes

Alkenes typically have chemical shifts between 4 and 6 ppm. Our alkene example will be a a vinyl group, which has some more interesting splittings. This group is in the compound acrylonitrile, which is shown in Figure 72.

The vinyl group is a good way to discuss a few details about alkene protons. First, the coupling constants are diagnostic for where they are with respect to each other. For example, the signals in acrylonitrile are the following: H_a is at 5.69 (1H, dd), H_b is at 6.11 (1H, dd), and H_c is 6.24 (1, dd). All three signals are doublets of doublets because all three protons are in different chemical environments. The two protons that are *cis* to one another (H_b and H_c) have a coupling constant of about 12 Hz. The two protons that are on the same carbon have a coupling constant of around 1 Hz. The remaining two protons trans to one another have a larger coupling constant of 18 Hz. These coupling constants are typical for protons when they are coupled to other protons on alkenes.

Alkynes

For alkyl alkynes, the terminal C-H proton is around 1.9 ppm, while for compounds like phenylacetylene, the terminal alkyne C-H proton is at around 3 ppm, shown in Figure 73.

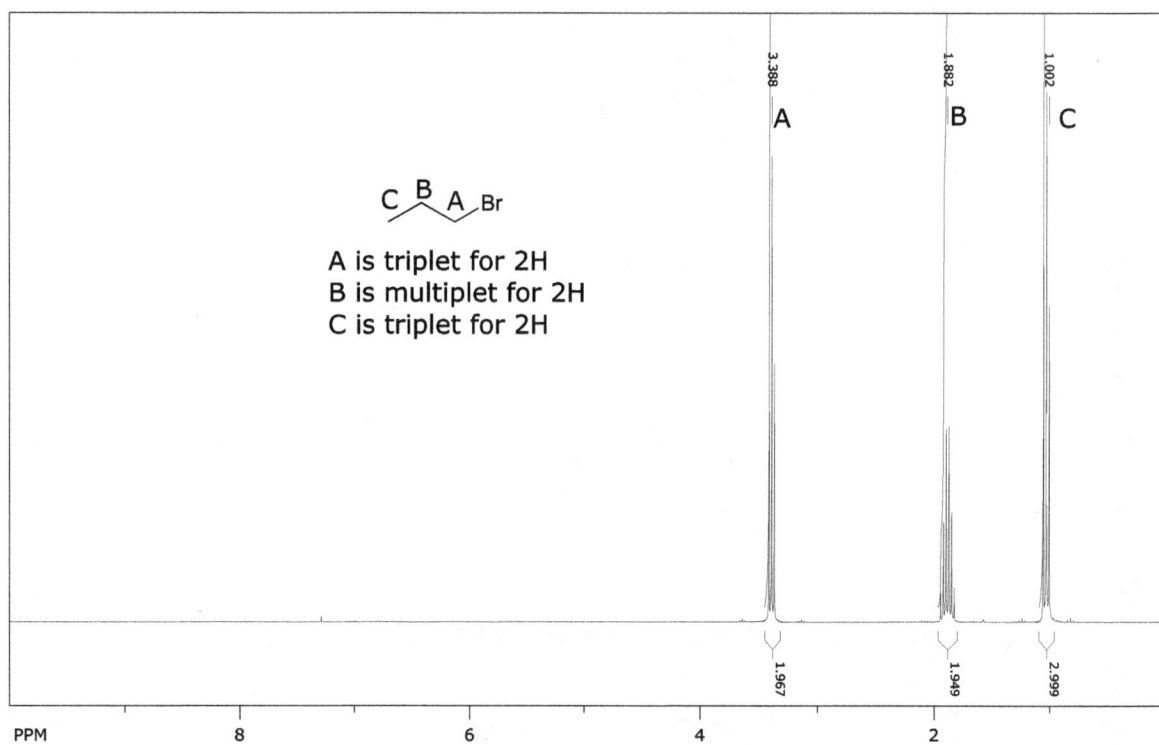

C B A Br

A is triplet for 2H
B is multiplet for 2H
C is triplet for 2H

ht!

Figure 71: ^{1}H NMR of 1-bromopropane

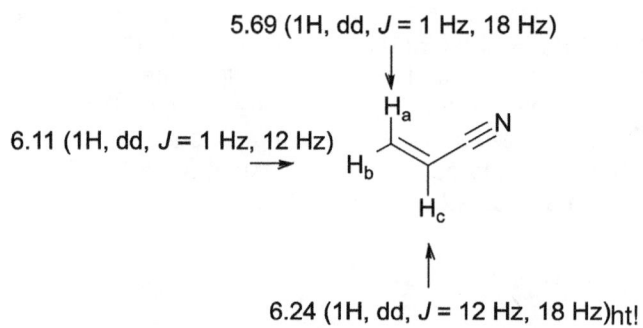

5.69 (1H, dd, J = 1 Hz, 18 Hz)

6.11 (1H, dd, J = 1 Hz, 12 Hz)

6.24 (1H, dd, J = 12 Hz, 18 Hz)ht!

Figure 72: ^{1}H NMR shifts for acrylonitrile

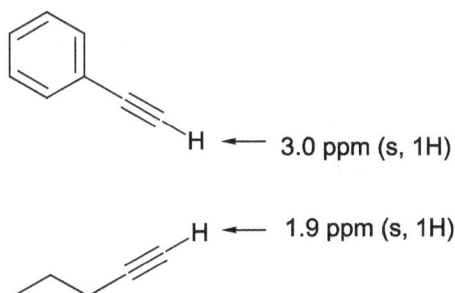

Figure 73: ^{1}H NMR shifts for terminal alkynes

Alkyl Halides

For alkyl halides, the proton shifts can be illustrated with the following examples: ethyl chloride, ethyl bromide, and ethyl iodide. Figure 74 shows that for each compound, the CH_2 next to the electronegative element is significantly further downfield than the CH_3 group.

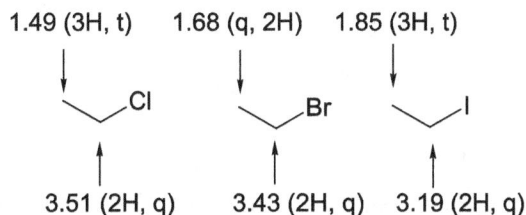

Figure 74: ^{1}H NMR shifts for alkyl halides

Aromatics

Aromatic compounds comprise a large number of structures; we will focus on mostly benzene derivatives, since each type of aromatic compound has its unique characteristics. Benzene itself only has one signal around 7.3 ppm due to the highly symmetrical structure. Unlike many other molecules we see in this chapter, delocalization/resonance can play a role in where the chemical shifts of protons occur due to the pi-system. Methoxybenzene and toluene will be a good place to start(Figure 75).

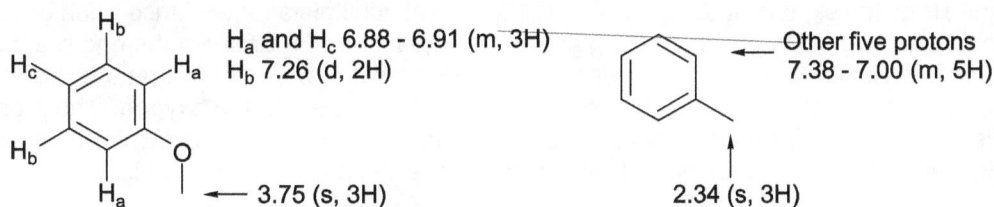

Figure 75: ^{1}H NMR shifts for aromatic compounds

The CH_3 of the methoxy group is a singlet at 3.75 ppm. This is a little farther downfield than some other ethers due to the aromatic ring. The aromatic protons upon first glance are two signals, although in reality there are three unique types of protons: two *ortho* to the methoxy, two *meta*, and one *para*. The two *ortho* to the methoxy are practically a doublet and are farthest upfield since they are closest to the methoxy. The para overlaps in part with the *ortho* signal, and is also practically a doublet (there is actually a small three-bond coupling between a proton and the proton *meta* to it).

Visually, these would show up as a multiplet for 3H at 6.9 ppm becuase of the overlap. The protons meta to the methoxy group are further downfield at 7.26 ppm and would be a doublet of doublets since there are the two different kinds of protons adjacent to them. In contrast, toluene is unable to donate via delocalization, and its spectrum is much simpler, with a 3H singlet at 2.34 ppm and a 5H multiplet at 7.38 to 7.00 ppm.

Another easily identified pattern in aromatic compounds are *para*-disubstituted aromatic systems. These can be identified by the two separate doublets in the aromatic region that each integrate for 2H. A good example of this sort of splitting is *p*-ethylbenzaldehyde. This compound has five signals as shown in Figure 76.

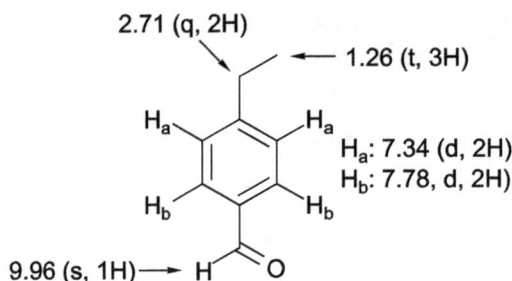

Figure 76: ^{1}H NMR shifts for *p*-ethylbenzaldehyde

The ethyl group on the ring is a typical ethyl group, with two signals: 2.71 (q, 2H) and the CH_3 group (1.26, t, 3H). The aromatic protons can be split into two sets of two due to symmetry: H_a at 7.34 (d, 2H) and H_b at 7.78 (d, 2H). H_b is farther downfield since it is next to the aldehyde C-H at 9.96 (s, 1H). Again, symmetry plays a role in the number and type of signals for this compound.

Alcohols and Phenols

An example of an alcohol and phenol can be seen in Figure 77. The alcohol in this case is butan-1-ol, which is a straight chain alcohol. The -OH proton is a broad singlet at 2.24 for 1H. As with other compounds, the chemical shift of the protons moves upfield with each carbon as we move away from the oxygen. Thus, the signals at 3.63, 1.53, and 1.39 are all multiplets for 2H since each of them have two different types of protons on neighboring atoms. As usual, the CH_3 group at the end is a triplet. The p-propylphenol is similar to the methoxybenzene compound we saw earlier in that the protons on the carbons *ortho* to the oxygen are much farther upfield than those *meta* to the oxygen. The propyl group attached is an isolated propyl group, with the CH_2 next to the aromatic ring being a triplet for 2H, the middle CH_2 being a multiplet, and the CH_3 at the end of the chain is a triplet for 3H.

Ethers

We will use diethyl ether as our ether example in Figure 78. Recall that ethers have the structure R-O-R, where R is generally an alkyl or aryl group.

0.94 (t, 3H)   1.53 (m, 2H)

2.51 (t, 2H) ←— 0.92 (t, 3H)

←— 1.59 (m, 2H)

OH ←— 2.24 (br s, 1H)

H_a   H_a

H_b   H_b

H_a: 7.04 (d, 2H)
H_b: 6.75,(d, 2H)

1.39 (m, 2H) 3.63 (m, 2H)

4.92 (s, 1H) —→ OH

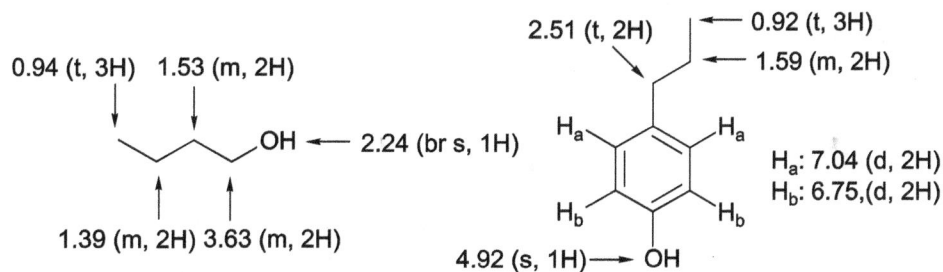

Figure 77: ^{1}H NMR shifts for alcohols and phenols

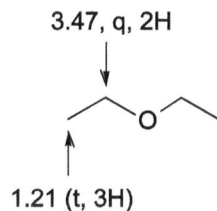

3.47, q, 2H

O

1.21 (t, 3H)

Figure 78: ^{1}H NMR shifts for diethyl ether

It has four carbons and ten protons, but only two signals in the ^{1}H NMR due to symmetry. Symmetrical molecules tend to have fewer signals than at first look, since the environments the protons find themselves in are the same for both sides during the experiment. Thus, we have the signals at 3.47 (q, 2H), and 1.21 (t, 3H). Similar to the alkyl halide example, the protons on the carbon adjacent to the oxygen are much further downfield because of the electronegativity of the carbon causing deshielding to occur.

Aldehydes and Ketones

Aldehydes have a distinct C-H proton shift of the H bound to the carbonyl (9 to 10 ppm), which makes them easy to identify. An example is shown in in Figure 79.

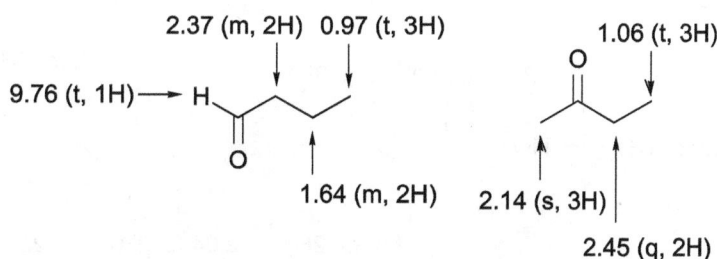

2.37 (m, 2H) 0.97 (t, 3H)

1.06 (t, 3H)

O

9.76 (t, 1H) —→ H

O

1.64 (m, 2H)   2.14 (s, 3H)

2.45 (q, 2H)

Figure 79: ^{1}H NMR shifts for aldehdyes and ketones

Butyraldehyde has four signals; the proton attached to the carbonyl has a chemical shift at 9.76 ppm and is observed as a triplet due to the CH_2 next door. The CH_2 and the CH_2 next to that are both multiplets at 2.37 and 1.64 ppm, respectively. The signal at 2.37 is a multiplet because it has two different neighbors; it can be easy to forget about the C-H attached to the carbonyl. Next we will look at an example of a ketone, butan-2-one. The structure of this compound contains an ethyl group and a methyl group. The methyl group is a singlet for 3H at 2.14, and the CH_2 of the ethyl group is a quartet for 2H at

2.45 ppm. Finally the CH_3 of the ethyl group is a triplet for 3H at 1.06 ppm.

Thiols and Sulfides

Thiols have an -SH group, and since S is more electronegative than C, the CH_2 next to the sulfur is far-thest downfield, and then as we move away from the S, the chemical shifts become farther and farther upfield. Protons on heteroatoms (like S), can behave in a variety of ways depending on the solvent, from not appearing in the NMR to being a broad signal to splitting like other protons. Figure 80 shows the shifts for propane-1-thiol and diisopropylsulfide.

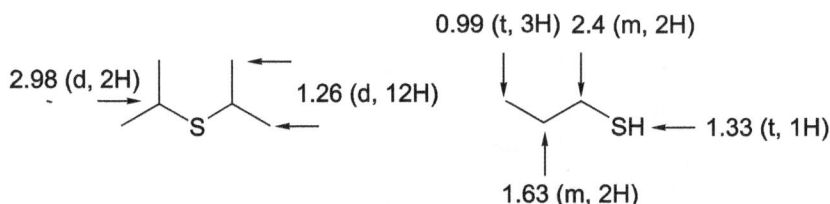

Figure 80: ^1H NMR shifts for thiols and sulfides

For propane-1-thiol, the proton on the sulfur is a triplet at 1.33 pm, the CH_2 next to the S is a multiplet at 2.4 ppm due to the -SH and the adjacent CH_2, the CH_2 adjacent to that is a multiplet at 1.63 ppm, and the CH_3 on the end is a triplet at 0.99 ppm. Diisopropyl sulfide has two signals, even though it contains 14 protons. There is a doublet for 12H at 1.26 ppm. This is from the four methyl groups being symmetrical. There is also a septet for 2H at 2.98 ppm from the two C-H protons.

Carboxylic Acids, Carboxylic Acid Derivatives, and Nitriles

Carboxylic acids, esters, amides, and nitriles are up next. Carboxylic acids have a distinct broad singlet, typically between 10 and 12 ppm from the -OH of the carboxylic acid group. First, let's look at butyric acid and ethyl acetate(Figure 81).

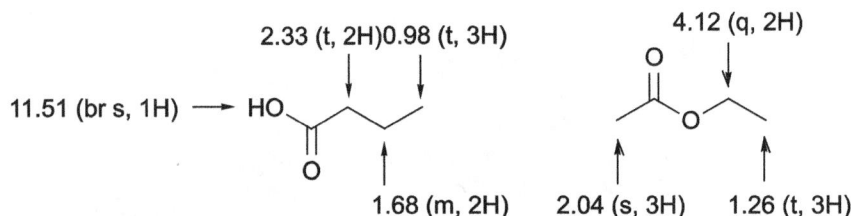

Figure 81: ^1H NMR shifts for carboxylic acids and esters

The -OH of the acid is a broad singlet at 11.51 ppm, while the other three signals are a triplet at 2.33, a multiplet at 1.68, and a triplet at 0.98 ppm. For ethyl acetate, we can see that there are three signals for three kinds of protons: the methyl directly attached to the carbonyl that is a singlet at 2.04 ppm, the methylene (CH_2) group attached directly to the oxygen at 4.12 ppm, and the methyl group attached to the methylene group at 1.26 ppm. The chemical shifts of the protons directly attached to the carbonyl and the oxygen are typical for an ester. In addition, we see two of the groups we previously talked about (methyl and ethyl).

Next, let's look at an amide and a nitrile, which can be seen in Figure 82.

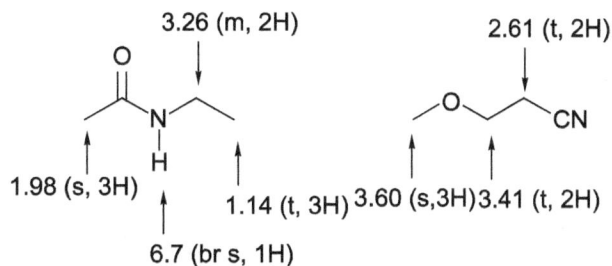

Figure 82: ^{1}H NMR shifts for amides and nitriles

For *N*-ethyl acetamide, the N-H proton is a broad singlet at 6.7, and the methylene adjacent to the nitrogen is a multiplet for 2H at 3.26 due to the N-H and the adjacent methyl group. The methyl group next to the methylene is a triplet for 3H, while the methyl next to the carbonyl is a singlet for 3H at 1.98 ppm. For 3-methoxypropionitrile, the methyl group next to the oxygen is a singlet for 3H since there are no neighboring atoms with protons, and the other two methylene groups are both triplets. The protons on the carbon attached to the oxygen is farther downfield at 3.41 ppm, while the protons on the carbon adjacent to the nitrile are at 2.61 ppm.

Amines

For our amine example, we will use 3-methoxypropylamine, which is a primary amine. We can see the structure and the assignments in Figure 83.

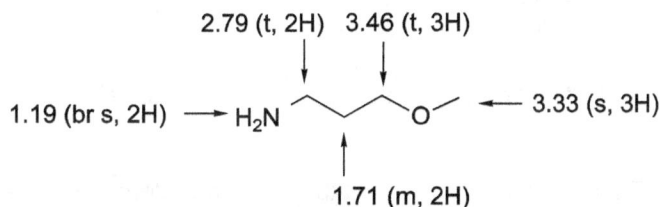

Figure 83: ^{1}H NMR shifts for 3-methoxypropylamine

The compound has five signals, four from each of the carbons, and then one for the -NH$_2$ group. The CH$_2$ group attached to the oxygen is a triplet for 2H at 3.45 ppm. The methyl group attached to the oxygen is a singlet for 3H at 3.33 ppm. Next is the -CH$_2$ group attached to the nitrogen atom, which shows up as a triplet for 2H at 2.79 ppm. The central CH$_2$ group is a multiplet at 1.71 ppm, and the NH$_2$ from the amine shows up as a broad singlet for 2H at 1.19 ppm.

Reactions

For organic chemistry, reactions are often at the heart of many questions, whether directly asked about or implied. Thus, the last two sections of this book are devoted to reactions (single steps) and syntheses (two or more reactions in a sequence). For each functional group we will first discuss reactions that generate that particular functional group, and then discuss reactions that use that functional group in the starting material. These sections will not cover every single possible reaction or synthesis, but it will aid you in thinking about reactions. There are three main parts to an individual reaction that are written out:

1. **Starting material.** This is the substrate we are beginning with. In general, this is the organic compound we are going to modify. There may be more than one listed with a + sign between them. Unfortunately, there are no strict guidelines for writing out reactions.

2. **Reagents/conditions/time/temperature.** Basically, this section is where the rest of the information that is given about the reaction goes. In general, the reagents go above the arrow, and the solvent, temperature, and time go below the arrow. Again, this is not set in stone, but is generally how it works. Sometimes, two or more steps are depicted here in abbreviated form, with the individual reactions/steps each having a number in front of them. The number indicates that the step is given time to complete.

3. **Product.** This is the desired/isolated organic product. There are usually other products that are not represented. Often, inorganic products are not written out. This is markedly different than what we often see in general chemistry where all products are given.

There are a number of different types of questions that can be asked, and we will go over some the most common.

1. **Predict the product.** These questions give you the starting material and the conditions, leaving you to give the product. These typically have only one correct answer.

2. **Give conditions.** These questions give you the starting material and the product, and ask for conditions. These can have more than one "correct" answer in the sense that more than one set of conditions can accomplish the same transformation. For example, reductions of aldehydes to alcohols can be accomplished with both $NaBH_4$ and LAH. Be sure to watch for modifiers/clues such as "best conditions." These can be multiple choice or open response.

3. **Give starting materials.** These questions give you the conditions and the product, and ask for conditions. These can also have more than one correct answer. For example, both primary alcohols and aldehydes will give carboxylic acids on reaction with chromic acid. Again, watch for modifiers/clues that point you toward a particular answer. These can be multiple choice or open response.

4. **Which of these . . .** These questions will compare two or more reactions and ask questions that

include but aren't limited to the following: Which reaction goes faster? Which reaction goes slower? Which reaction gives a higher yield of desired product? Which reaction will give a meso product? Which reaction will give a chiral product? Here, the key is to be able to know enough about the reaction mechanism to find the correct answer. These can be multiple choice or open response.

5. **What is true/false about . . .** These questions will test what you know about mechanisms or reaction outcomes. Examples include but aren't limited to the following: Which statement about alkyne reactions are true? Which statement about alkene oxidations are false? These tend to be multiple choice.

Finally, when we are looking at the reactions, a functional group that is a product of one reaction may be seen as a starting material for a different reaction. Thus, reactions will not be repeated, but referred to if they have already been discussed.

Alkanes

Alkanes as starting materials. Alkanes are not regularly used as reactants in organic chemistry. One type of reaction that uses alkanes is the radical halogenation reaction. Examples of questions involving this are shown. The first is a typical substitution of an alkane (2-methylpropane) using bromine and light and can be seen in Figure 84. This is a predict the product question.

What are the possible products of this reaction?

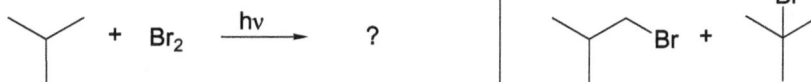

Figure 84: Radical bromination of 2-methylpropane

To the right of the vertical line we see the two main products: 1-bromo-2-methyl propane on the left and 2-bromo-2-methyl propane on the right. This question is slightly different than the related question, "What would be the major product of the reaction?" That question is less intuitive to answer since there are nine primary hydrogens in the starting material and one tertiary hydrogen. Yet, it is known that the tertiary hydrogen is much more easily substituted. Thus, the product on the right is likely to be the *major* product of the reaction.

Alkanes as products. There are a number of reactions that give alkanes as products. These include the reduction of alkynes and alkenes and the reduction of aldehydes or ketones by reaction with hydrazine and then KOH (Wolff-Kishner reduction). Examples of these are shown in Figure 85.

In (a), we can see that we have a question that asks us to predict the product of the reaction of cyclohexene with hydrogen and palladium on carbon catalyst. Alkenes are reduced to alkanes under these conditions, so the answer is cyclohexane. The hydrogen atoms add to the same face, giving a *syn*-addition. At the end of this section in the practice problems you will see the impact of this. In (b), we see the reaction of 1-butyne to gave butane, and the question asks which conditions are best. Here, since we are going from alkyne to alkane, the hydrogenation conditions here are best. The dissolving metal (Na/NH_3) or Lindlar/deactivated catalyst type conditions won't work since they only stop at the alkene. In (c), we have a set of conditions and a product, and the question wants to know what starting material could be used in the reaction. Here, there is more than one answer. The Wolff-Kishner reduction converts an aldehyde or ketone to an alkane. Thus, if we have an alkane for a product, there may be

(a) What is the major product of this reaction?

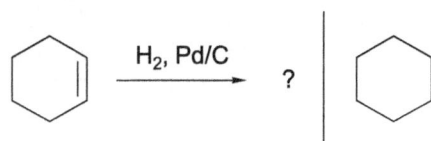

(b) What conditions are best to generate the product?

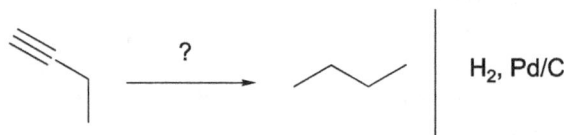

(c) What starting material could be used in the following reaction?

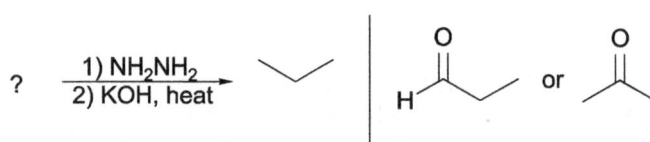

Figure 85: Alkanes as products of various reactions

multiple answers, like we see here. Depending on symmetry or other substitutions present, there may be only one answer.

Alkenes

Alkenes as starting materials. Alkenes are common feedstock chemicals and are often used as starting materials for a variety of reactions. Most (but not all!) alkene reactions involve breaking the π-bond and converting into two σ-bonds. There are a large number of them, and they will be discussed.

Our first three examples are shown in Figure 86.
In (a), we have a question that asks about which of the products of the reaction are meso. Recall that meso compounds have stereocenters but also an internal plane of symmmetry so that overall the molecule is still achiral. Thus, answer B is the only one that gives a meso compound. As we saw previously in the book, if we rotate around the central C-C bond of answer B, we will see the mirror plane directly.

In (b), we have a similar question, but now an alkene reacting with D_2 instead of H_2. Since deuterium is an isotope of hydrogen, it is different enough that it will generate up to two stereocenters in the products. Here, we see answers A nd C are meso products since an internal plane of symmetry can be drawn between them. For (c), we have a relatively simple question of HBr reacting with an alkene. This reaction is going to follow Markovnikov's rule and give the product shown, where the -Br ends up on the more substituted carbon of the reactant. Next, we will look at another group of alkene reactions with typical questions, which can be seen in Figure 87.

In (a), we have a question that is asking for a suitable starting material for the reaction shown. Basically, we know that our product is a bromohydrin, and so the thought process would be to look back and think of what alkene would be a suitable starting material. In (b), we have a question about conditions. The reaction is converting an unsymmetrical alkene (notice how one end has one carbon attached and

(a) Which of these alkenes will produce a meso product when reacted with Br$_2$?

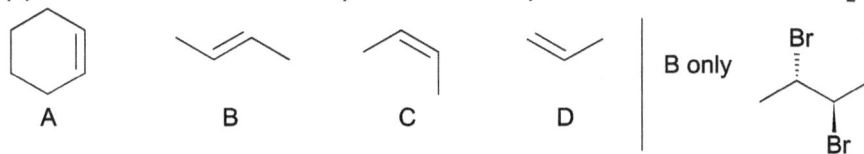

A   B   C   D | B only

(b) Which of these alkenes will produce a meso product when reacted with D$_2$, Pd/C?

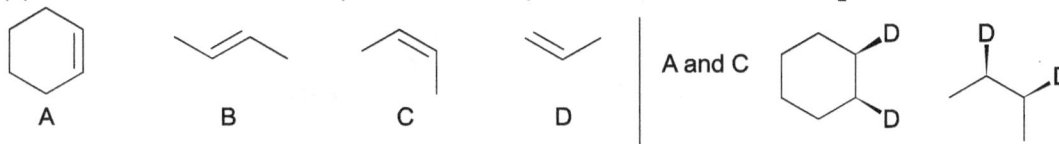

A   B   C   D | A and C

(c) What is the product of the following reaction?

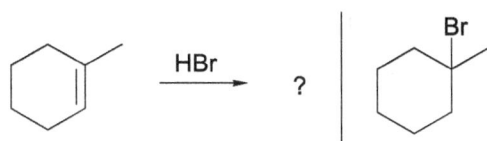

HBr → ?

Figure 86: Alkenes as starting materials in reactions

(a) What is the best starting material for this reaction?

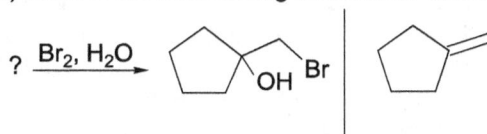

? $\xrightarrow{Br_2, H_2O}$

(b) Which conditions are best for this transformation?

? →

1) BH$_3$ (or other derivative)
2) NaOH, H$_2$O$_2$

(c) What is the product of this reaction?

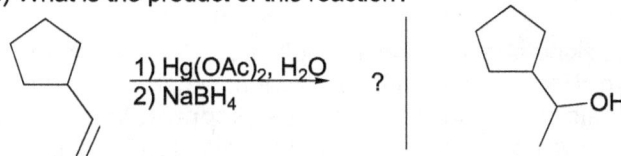

1) Hg(OAc)$_2$, H$_2$O
2) NaBH$_4$?

Figure 87: Examples of alkene reactions

the other has no carbons) into a primary alcohol. The recognition of it being a primary alcohol is key to the answer, as we see how the -OH is on the less substituted side. This incidates that it is best synthesized by doing a hydroboration-oxidation reaction. In (c), we have a similar reaction, but wanting a product. Here we also have an unsymmetrical alkene. The reaction conditions indicate it is a oxymercuration-demercuration reaction. This leads to the -OH ending up on the more substituted carbon of the alkene. Let's look at another set of reactions involving alkenes as starting materials. These can be seen in Figure 88.

(a) Which conditions are best for the following transformation?

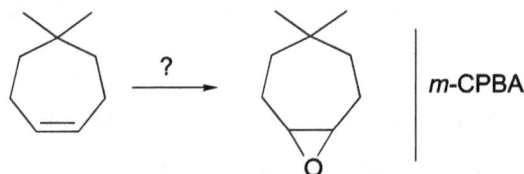

m-CPBA

(b) What is a good starting material for this transformation?

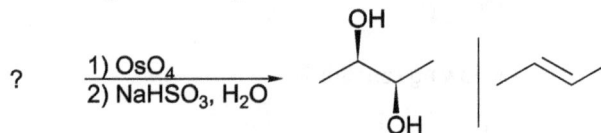

1) OsO$_4$
2) NaHSO$_3$, H$_2$O

(c) What is the product of the following reaction?

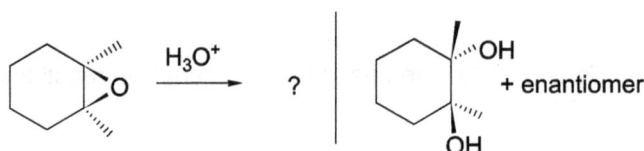

H$_3$O$^+$

? + enantiomer

Figure 88: Reactions of alkenes and an epoxide

In (a), we have a question about best conditions for an epoxidation. While there is more than one option, *m*-CPBA is a very common reagent for this purpose. In (b), we have a question about what a good starting material is for the following reaction. One key feature to notice is that we have reaction conditions that point to a syn-dihydroxylation, and a product that is drawn such that the -OH groups are cis. Be on the lookout for questions that have the -OH groups trans to one another. Finally, we have a reaction in (c) where we have the reaction of an epoxide in acidic water. The end result is a *trans*-disposition of the two -OH groups in the product. Here, we have a cyclic system, so it is easily seen as one -OH is denoted with a wedge, and the other -OH with a dash. A further note is that in most undergraduate texts, the products are racemic, and thus could be written as it is here, + enantiomer. Be sure to check whether you need to make that explicit. Now let's look at our next to last set of alkene reactions in Figure 89.

In (a), we have a question asking about the best conditions for converting the cycloheptene into the cyclopropane derivative shown. Here, the key is to realize that we are making a cyclopropane from an alkene. Thus, the Simmons-Smith carbenoid route is most appropriate. In (b) we have a similar reaction, except we are making a dichlorocyclopropane. Thus, we are going to start with an alkene. Recognizing that this process is concerted leads to the correct answer of cis-2-butene. Finally, let's look at our last set of reactions involving alkenes as starting materials. The reactions are shown in Figure 90.

In (a), we have a cyclopentene being converted into a diketone. Notice how no carbons are added or removed. The double-bond is broken and converted into two carbonyls. Since that is the case, ozonolysis is a good answer. For ozonolysis reactions, the first step is always the same, with O$_3$ always first. However, depending on the substrate and the desired outcome, other reagents/conditions can be seen in step 2. For example, dimethylsulfide and triphenylphosphine are common reducing agents used in the second step. If the alkene has a Csp^2-H bond on the alkene, the reducing conditions are necessary to keep the reaction from oxidzing to the carboxylic acid. In (b), we see an example of oxidative conditions

(a) Which conditions are best for the following transformation?

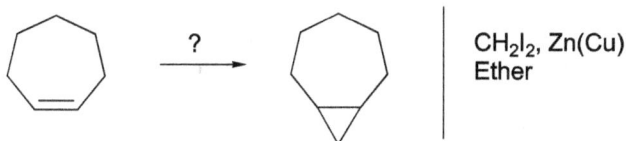

CH$_2$I$_2$, Zn(Cu)
Ether

(b) What is a good starting material for this transformation?

?   KOH / CHCl$_3$ →

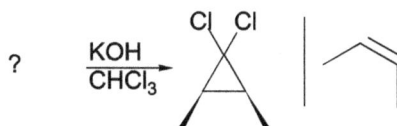

Figure 89: Cyclopropanation reactions of alkenes

(a) Which conditions are best for the following transformation?

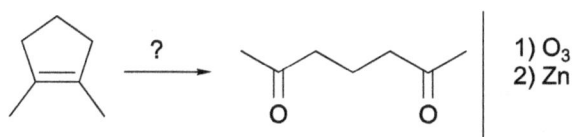

1) O$_3$
2) Zn

(b) What is the product of this reaction?

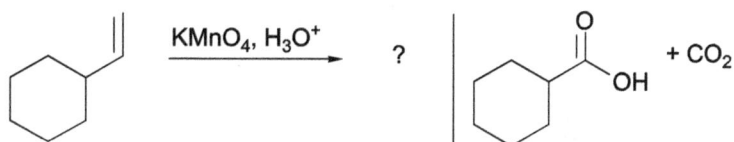

KMnO$_4$, H$_3$O$^+$ → ?

+ CO$_2$

Figure 90: Alkene cleavage reactions

that DO convert the Csp^2-H bonds an the alkene into C-O bonds. The acidic permanganate will convert the C=C into two C=O like we saw earlier, but these conditions will also convert those C-H bonds into C-O bonds. CO$_2$ is generated as a product because the reactant is a terminal alkene. Here we have seen that alkenes can react under a wide variety of conditions to give a wide range of product structures.

Alkenes as products. There are a number of ways to generate alkenes as products, including the reaction of aldehydes or ketones with phosphorus ylides (Wittig reaction), reductions of alkynes, and various elimination reactions. Our first three examples are shown in Figure 91:

In Figure 91(a) we see a question asking about the conversion of a ketone into an alkene. While there are other reagent combinations that can perform this transformation, the Wittig olefination is among the best. Notice how the three carbons added to the ketone come from the phosphorus reagent. In (b), we predict the product question about converting an internal alkyne into a cis-alkene. Lindlar's catalyst (or other modified or poisoned catalysts) stop at one reduction. In (c), we see a question about best conditions. Here we see the conversion of an internal alkyne into a trans-alkene. For this transformation, the best conditions are a metal such as sodium with liquid ammonia as the solvent (dissolving metal reduction). Next, let's look at some eliminations that give alkenes in Figure 92.

In Figure 92(a) we see we have a question about predicting the product of a reaction involving a secondary bromide and potassium ethoxide in ethanol. These conditions point toward elimination, so

(a) What is the best reagent for this transformation?

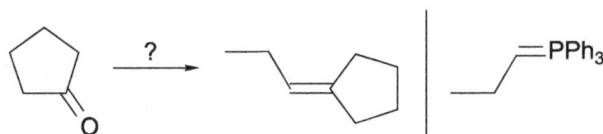

(b) What is the product of the following reaction?

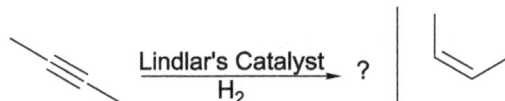

(c) Which conditions are best for the following transformation?

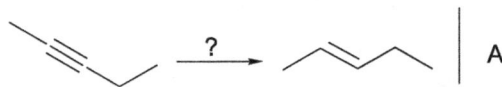

A) Na, NH₃ B) Lindlar/H₂ C) H₂, Pd/C D) HCl

Figure 91: Reactions to give alkenes

(a) What is the product of this reaction?

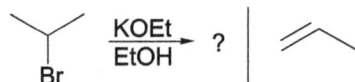

(b) What conditions are best for this transformation?

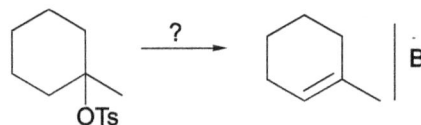

A) HCl B) NaOEt C) NaCl D) EtOH

(c) What is a good starting material for this reaction?

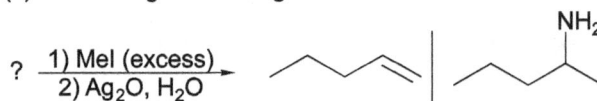

Figure 92: More reactions to give alkenes

our product is propene. In (b), we have a tertiary tosylate being converted to a trisubtituted alkene. Of the options, B is the best. Choice A is acidic, and C and D are neutral, none of which would lead to predominant elimination. In (c), we see we have a two-step conversion to a terminal alkene. In particular, we see that the first step is excess methyl iodide, followed by silver oxide. These conditions (when combined with the amine shown) are named the Hofmann elimination. This gives the *less* stable alkene product.

Alkynes

Alkynes as starting materials. Alkynes are broadly similar to alkenes in that they have π-bonds. Depending on the conditions, one or both π-bonds may be reacted. Reactions include reductions, hydrations, and additions of HX and X_2. Terminal alkynes and acetylene can also be deprotonated and used as a nucleophile in S_N2 reactions. Reductions of alkynes to alkenes and alkanes were shown in Figure 91. Some hydration reactions of alkynes are shown below in Figure 93.

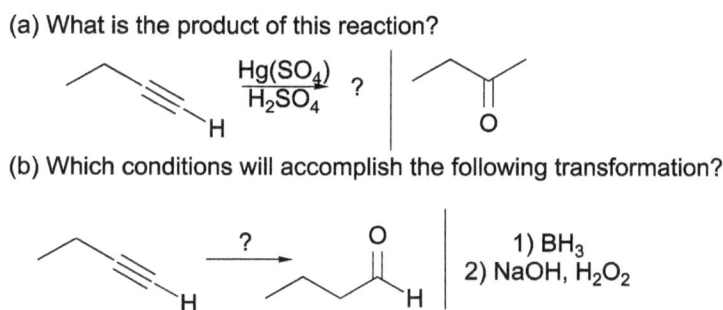

(a) What is the product of this reaction?

(b) Which conditions will accomplish the following transformation?

Figure 93: Hydrations of terminal alkynes

Figure 93(a) shows a common question involving hydrations of terminal alkynes. Here, we have a terminal alkyne reacting with a mercury salt. This will always give a methyl ketone if the starting material is a terminal alkyne. The other example is a question about reaction conditions that give an aldehyde when starting from a terminal alkyne. Here is the hydroboration-oxidation reaction. Note that there are other borane derivatives (not just BH_3) can be used here as well. Next we will discuss additions of HX and X_2 to alkynes. Typically, two equivalents of HX and X_2 involve two equivalents to end up with saturated products. Some versions of questions in Figure 94.

(a) What is the product generated when one equivalent of bromine is reacted with 2-butyne? When two equivalents are reacted?

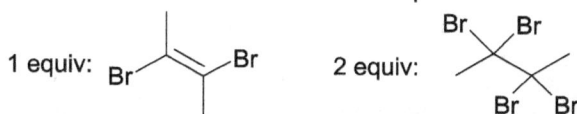

(b) What is the product generated when one equivalent of HBr is reacted with 1-butyne? When two equivalents are reacted?

Figure 94: Reactions of alkynes with Br_2 and HBr

Alkynes react similarly to alkenes in that for (a) we see that one equivalent of bromine reacting with 2-butyne gives (2E)-2,3-dibromobut-2-ene and two equivalents gives 2,2,3,3-tetrabromobutane. In (b), we see that Markovnikov's rule is followed where the reaction puts the Br on the more substituted carbon of the alkyne, and with one equivalent of bromine reacting we observe 2-bromobut-1-ene as the product. When two equivalents are added, we observe 2,2-dibromobutane as the product since the Br can also stabilize the carbocation via resonance/delocalization.

Alkynes as products. As mentioned, terminal alkynes and acetylene can be deprotonated and used as nucleophiles. These reactions also result in alkyne products. Also, alkynes can be generated by elimination reactions. Examples are shown in Figure 95.

In Figure 95(a), we have a question about a starting material. The conditions ($NaNH_2$, then an alkyl halide) are typical for using alkynes to make C-C bonds. It is important to remember that the pKa of the

(a) What is the starting material for this reaction?

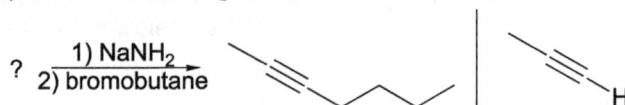

$$? \quad \xrightarrow[\text{2) bromobutane}]{\text{1) NaNH}_2}$$

(b) What is the product of this reaction?

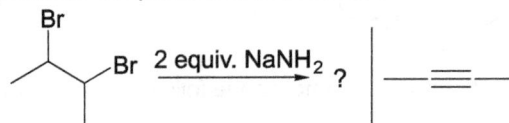

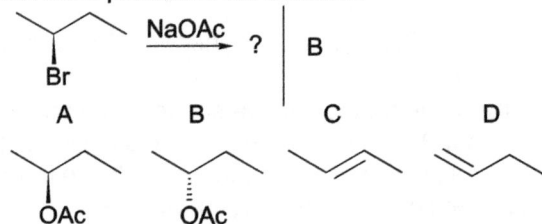

Br

$$\xrightarrow{\text{2 equiv. NaNH}_2} ?$$

(c) What is the product of this reaction?

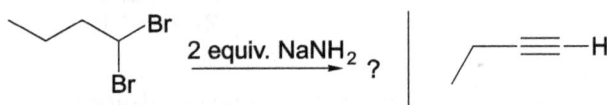

Br

$$\xrightarrow{\text{2 equiv. NaNH}_2} ?$$

Br

Figure 95: Reactions affording alkynes as products

terminal alkyne Csp-H bond is around 25, so a strong base is needed. Besides sodium amide, butyl lithium (BuLi) and Grignard reagents can also be used to generate the acetylide (deprotonated acetylene). The electrophile is limited to methyl halides (methyl iodide, for instance) or primary alkyl halides.

In (b) and (c) we have a vicinal dihalide (2,3-dibromobutane) and a geminal dihalide (1,1-dibromobutane) reacting to give alkynes. Either substitution pattern can work since they both give a vinyl halide intermediate, which eliminates the alkyne.

Alkyl Halides

Alkyl Halides as starting materials. Alkyl halides are common reactants, and their use is generally split into two main categories: (1) substrates for subsitution reactions and (2) substrates for elimination reactions to give alkenes or alkynes. An example of a substitution is shown in Figure 96; their use as substrates to give alkenes and alkynes shown in Figure 92 and Figure 95.

What is the product of this reaction?

$$\xrightarrow{\text{NaOAc}} ? \quad \text{B}$$

Br

A   B   C   D

OAc   OAc

Figure 96: An S$_N$2 reaction of an alkyl halide

In Figure 96, we see that we have a question about the product of a reaction of (2S)-2-bromobutane with sodium acetate. The answer is B since the sodium acetate is going to act is a nucleophile and not a base in this case. Thus, the S$_N$2 reaction will occur, giving (2R)-butan-2-yl acetate as the product. Next, let's look at some reactions that give alkyl halides as products.

Alkyl halides as products. There are a few types of reactions that give alkyl halides as products: (1) conversion of alcohols using $SOCl_2$ or HCl, (2) radical substitution of hydrocarbons with a halogen, and (3) reaction of alkenes with HX. Examples are shown in Figure 97.

(a) What is the best reagent for this transformation?

A) KCl   B) Cl_2 C) $SOCl_2$ D) HCl

(b) What is the product of this reaction?

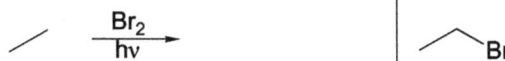

(c) What starting material would be useful for this reaction?

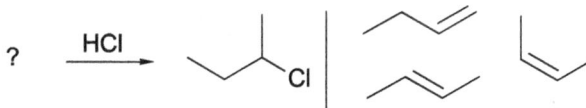

Figure 97: Reactions giving alkyl halides as products

In (a), the best reagent for converting propanol to propyl chloride is thionyl chloride ($SOCl_2$). KCl is unlikely to react, while Cl_2 and HCl would either be too reactive or give signficant side reactions. In (b), we have the reaction of ethane with elemental bromine in the presence of light. In ethane, all six hydrogens have equal reactivity, so the product will be bromoethane. In (c), there are a number of potential alkenes as starting materials, in this case, cis and trans-2-butene, as well as 1-butene. All three of these lead to the same carbocation intermediate, and thus the same product.

Aromatics

Aromatics as starting materials. Aromatic compounds will undergo three main types of reactions: (1) Electrophilic aromatic substitution, (2) Nucleophilic aromatic substitution, and (3) reactions adjacent to the ring at the benzylic position. Examples of these are shown in Figure 98. In (a), we have *p*-xylene reacting under typical nitration conditions (HNO_3 and H_2SO_4). Since the starting material is symmetrical, there is only one product generated, in this case 1,4-dimethyl-2-nitrobenzene. In (b), we have an example of a nucleophilic aromatic substitution with *p*-chloro nitrobenzene reacting with sodium ethoxide. The product is a substitution of the chloride with the nucleophile to give 1-ethoxy-4-nitrobenzene. Finally, in (c) we have a benzylic oxidation of ethylbenzene to benzoic acid using potassium permanganate. If there is a benzylic C-H bond, the permanganate will oxidize the compound to the carboxylic acid.

Aromatics as products of reactions. All of the reaction types shown (where aromatics are starting materials) will also give aromatic compounds as products. Reactions that produce aromatic compounds from non-aromatic compounds are not commonly seen in the undergraduate course sequence.

(a) What is the product of this reaction?

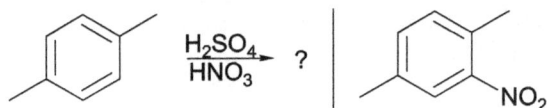

(b) What is the product of this reaction?

(c) What is the product of this reaction?

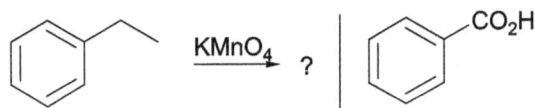

Figure 98: Examples of reactions of aromatic compounds

Alcohols and Phenols

Alcohols and phenols as starting materials. Alcohols are very common starting materials and can be converted into a number of products, including alkyl halides, ethers, esters, aldehydes, ketones, carboxylic acids, and alkenes. This is not an exhaustive list, but you can see how versatile they are. Phenols are compounds that have an -OH attached to a phenyl ring. These have distinctly different reactivity, and beyond making ethers aren't often seen as reactants at the undergraduate level. The first trio is shown below in Figure 99.

(a) What is the best reagent for this transformation?

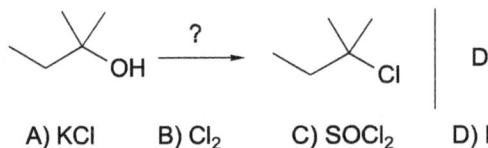

A) KCl   B) Cl_2 C) $SOCl_2$ D) HCl

(b) What conditions are best for this transformation?

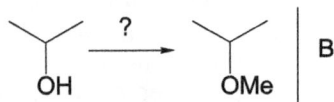

A) 1) HCl   B) 1) NaH   C) 1) MeMgBr   D) MeOH
   2) MeI   2) MeI   2) H_3O^+

(c) What is the product of this reaction?

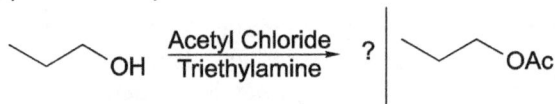

Figure 99: Reactions utilizing alcohols as starting materials

In (a), we are looking at converting a tertiary alcohol into the alkyl chloride. The best choice for this conversion is answer D. The tertiary alcohol will be protonated by HCl to generate the tertiary carbocation, which the chloride recombines with to give the product shown. In (b), we are looking to convert a secondary alcohol into an ether. Here, the best conditions are B. Conditions A would likely lead to elimination, while C would have step (1) deprotonate and then step (2) reprotonate, giving the starting material as the product. Conditions D would result in no reaction. Finally, in (c), we are reacting propanol with acetyl chloride under basic conditions (triethylamine). In this case, we will generate the ester propyl acetate.

Let's look at our next trio in Figure 100. In (a), we are looking to oxidize the primary alcohol propanol to the aldehyde propanal. Conditions A, C, and D will all take oxidize to propanoic acid, while conditions B is the only ones that will only oxidize to the aldehyde. In (b), we see the secondary alcohol being oxidized by chromic acid. For secondary alcohols, all of the typical alcohol oxidation conditions (chromic acid, PDC, PCC, etc.) will give the ketone since ketones cannot easily be further oxidized.

(a) What is the best reagent for this transformation?

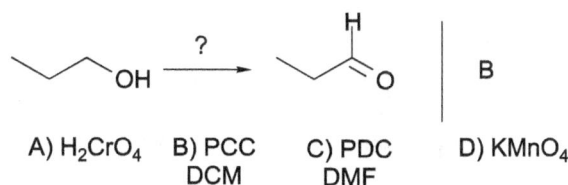

A) H_2CrO_4 B) PCC DCM   C) PDC DMF   D) $KMnO_4$

(b) What is the product of this reaction?

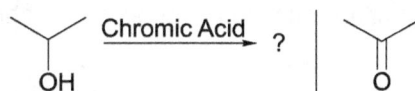

Chromic Acid

(c) What is the product of this reaction?

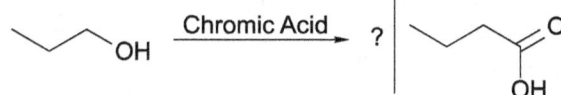

Chromic Acid

Figure 100: Alcohol oxidation reactions

Finally, let's look at a way to dehydrate alcohols to alkenes, shown in Figure 101. Here, we have a general question of how to convert alcohols to alkenes and conditions D are best. Conditions A, B, and C are unlikely to lead to conversion to alkenes.

Which conditions are best for the dehydration of alcohols to alkenes?

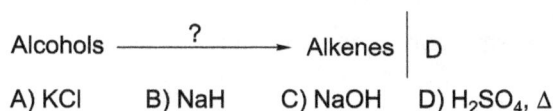

Alcohols ——?—→ Alkenes | D

A) KCl   B) NaH   C) NaOH   D) H_2SO_4, Δ

Figure 101: Dehydration of alcohols to alkenes

Alcohols and phenols as products. Alcohols are also very common products of reactions and can be derived from carboxylic acids, esters, aldehydes/ketones, alkyl halides, and alkenes. This is not an exhaustive list, but as we saw, alcohols are very useful compounds. Phenols are not often seen as products except as noted.

Let's look at our first set of reactions that generate alcohols, shown in Figure 102. In (a), we are asking what the product is when propanoic acid is reacted with lithium aluminum hydride (LAH), followed by acidic workup. LAH is known to reduce many kinds of carbonyl-containing compounds, and when the compound is a carboxylic acid, a primary alcohol is generated. In (b), the question is which conditions will reduce the ester to the primary alcohol. Here, the sodium borohydride in conditions A are not reactive enough to reduce the ester. Conditions C and D would give different products. Thus, the best conditions are those in B. In (c), 2-butanone is the best choice, reduced by $NaBH_4$ to give 2-butanol.

(a) What is the product of this reaction?

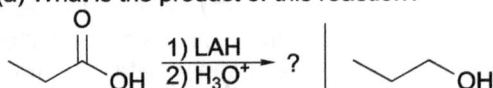

(b) What reaction conditions are best for this transformation?

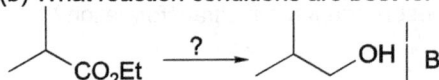

A) $NaBH_4$ B) 1) LAH   C) NaOH   D) H_2SO_4, Δ
            2) H_3O^+

(c) What starting material would be best for this reaction?

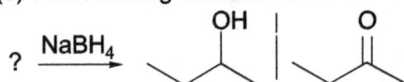

Figure 102: Reduction reactions to generate alcohols

Next, let's look at the reactions of alkenes that give alcohols as products in Figure 103. In (a), we are converting a terminal alkene into a secondary alcohol. The addition of water follows Markovnikov's rule, so the best set of conditions is the oxymercuration/demercuration sequence. In (b), the same alkene undergoes the hydroboration/oxidation sequence, so our product is one where the water is added in an anti-Markovnikov fashion, in this case the primary alcohol.

(a) What are the best conditions for this reaction?

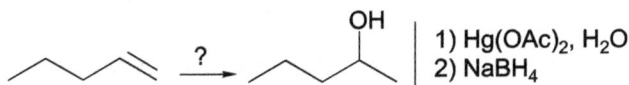

1) $Hg(OAc)_2$, H_2O
2) $NaBH_4$

(b) What is the product of this reaction?

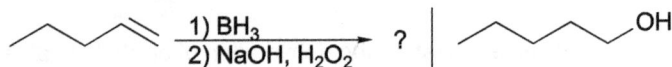

1) BH_3
2) NaOH, H_2O_2

Figure 103: Formal hydration reactions of alkenes

Ethers

Ethers as starting materials. Non-expoxide ethers are not often seen as starting materials since they are relatively stable compounds. Epoxides, on the other hand, are often used as starting materials since they have a three-membered oxygen containing ring that can be opened under acidic or basic conditions.

Examples are shown in Figure 104. In (a), we have a strongly nucleophilic and basic Grignard reagent reacting with an expoxide. In these cases, the strongly basic nucleophile will attack the less-hindered side of the epoxide, giving the product shown. In (b), we have acidic conditions and a relatively poor nucleophile in methanol. In this case, the nucleophile (methanol) will attack the more hindered side, giving the product shown with inversion at the stereocenter. In cases of epoxide reactions with strong acids, mixtures of attack at each carbon of the epoxide ring can result.

(a) What is the product of this reaction?

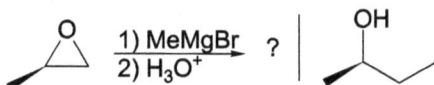

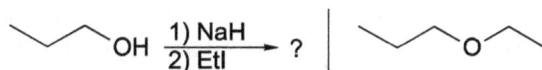

1) MeMgBr
2) H_3O^+

(b) What is the product of this reaction?

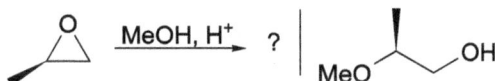

MeOH, H^+

Figure 104: Epoxide opening reactions

Ethers as products. Both non-epoxide ethers and other ethers are often seen as products in reactions. Non-epoxide ethers can be generally made by an S_N2 reaction under basic conditions, or by a more S_N1-type reaction under acidic conditions or in the presence of a Lewis-acidic ion like mercury. Examples of these are shown in Figure 105.

(a) What is the product of this reaction?

1) NaH
2) EtI

(b) What reaction conditions are best for this transformation?

D

A) MeOH   B) NaOMe   C) NaOH   D) H_2SO_4, Δ

(c) What starting material would be best for this reaction?

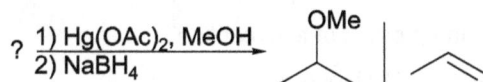

1) Hg(OAc)$_2$, MeOH
2) NaBH$_4$

Figure 105: Reactions that produce ethers

In (a), we see a traditional ether-forming reaction starting from a primary alcohol. The sodium hydride deprotonates the alcohol to give a good nucleophile, followed by reaction with ethyl iodide. An S_N2 reaction ensues, giving the ether product shown. In (b), we have an example of tertiary bromide being converted into an ether. Conditions B, C, and D are likely to result in elimination, while conditions A are likely to give the desired product via an S_N1 reaction. In (c), we have a modification of the oxymercuration conditions where methanol is used instead of water. Thus, the product is a methyl ether instead of an alcohol. Once we consider that propene is the best starting material for the reaction.

Thiols and Sulfides

Thiols and sulfides as starting materials. Thiols and sulfides are more commonly seen in biology and biochemistry but are covered in most organic textbooks. They are, broadly speaking, the sulfur analogues to alcohols and ethers. There are a few differences between the two groups, resulting from the larger sulfur atom. (1) Sulfur is more nucleophilic than oxygen, so substitution reactions happen more easily, (2) thiols are more acidic than alcohols, and (3) sulfur has more oxidation states available, so there are other accessible functional groups for sulfur, such as sulfoxides, and sulfones that do not have oxygen analogues. We will see these products, as well as their structures in Figure 106.

(a) What is the product of this reaction?

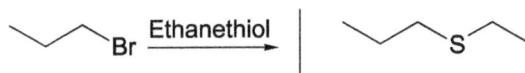

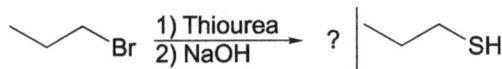

(b) What is the product of this reaction?

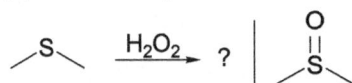

(c) What is the product of this reaction?

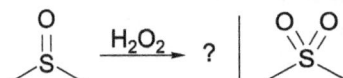

Figure 106: Examples of reactions with sulfur compounds

In (a), since we know that thiols can react similarly to alcohols, we imagine a version of the etherification we saw in Figure 105. Since sulfur is more nucleophilic than oxygen, we don't need to deprotonate first. Since sulfur is on the third row of the periodic table, in (b) and (c) we observe oxidations that we do not see with ethers. A variety of oxidants can be used for these reactions, not just H_2O_2. The product in (b) is a sulfoxide, and in (c) a sulfone.

Thiols and sulfides as products. Since sulfur has a more nucleophilic nature than oxygen, the reactions that produce thiols and sulfides will look slightly different. One reaction in particular is the use of thiourea to make sulfides due to the fact that sulfur is more nucleophilic than oxygen. This reaction and two others can be shown in Figure 107.

(a) What is the product of this reaction?

(b) What conditions are best for this reaction?

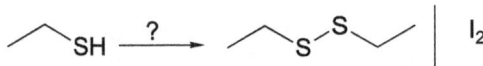

(c) What conditions are best for this reaction?

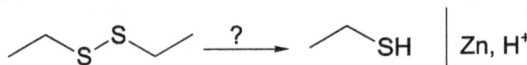

Figure 107: Reactions giving thiols and disulfides

In (a), we use thiourea as a nucleophilic sulfur source. Since thiourea already has a double bond to sulfur, we don't need to worry about it reacting more than it should. The second step is to hydrolyze the release of the thiol product. In (b), we see that we are actually starting with a thiol, but the reaction is placed here, since it works with (c) to show how thiols can be oxidized to disulfide (S-S bond), or a disulfide can be reduced to the thiol. In (b), we are oxidizing to the disulfide with iodine, while in (c), we are reducing the disulfide to a thiol with zinc and a strong acid.

Aldehydes and Ketones

Aldehydes and ketones as starting materials. Aldehydes and ketones are versatile starting materials that can give different types of products depending on reaction conditions. Aldehydes can be converted into alcohols, carboxylic acids, hydrazones, oximes, imines, and ketals. Ketones can be converted into alcohols, hydrazones, oximes, imines, and ketals. Let's look at our four in Figure 108.

(a) What starting material would be best for this reaction?

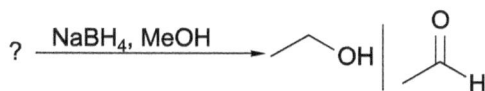

(b) What is the product of this reaction?

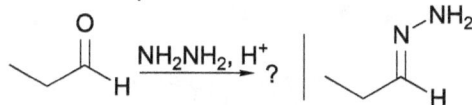

(c) What is the product of this reaction?

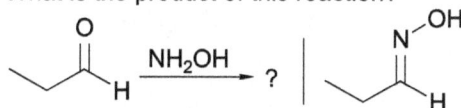

(d) What is the product of this reaction?

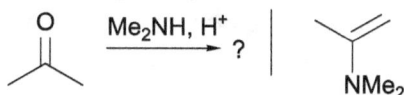

Figure 108: Reactions of aldehydes and ketones

In (a), we are asked to see which starting material would give ethanol upon reduction with sodium borohydride in methanol. Here, the answer is going to be ethanal, since we know that sodium borohydride cannot reduce esters. In (b) and (c), we are converting aldehydes into hydrazones and oximes, respectively. These are both condensation reactions, giving off water. Finally in (d), we have a secondary amine reacting with a ketone to give the enamine product shown. Next, let's look at some other reactions in Figure 109.

In (a), we see a reaction of an aldehyde with a primary amine to give an imine, giving water as the byproduct. In (b), we see the conversion of an aldehyde to an acetal using methanol under acidic conditions. For (c), we see a cylic acetal being formed. Acetals function as common protecting groups for ketones and aldehydes, with cyclic acetals being more stable than acylic acetals.

Aldehydes and ketones as products. Aldehydes and ketones can both be derived from alcohols, alkenes, and alkynes. Refer to Figure 100, Figure 90, and Figure 93 for these reactions.

(a) What is the product of this reaction?

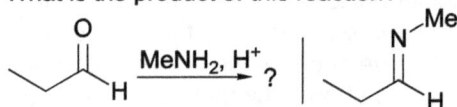

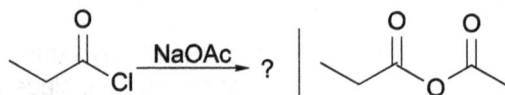

(b) What is the product of this reaction?

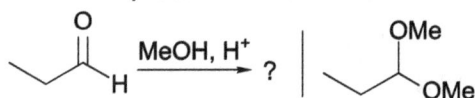

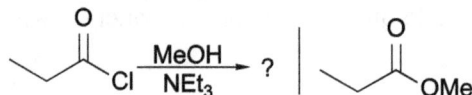

(c) What is the product of this reaction?

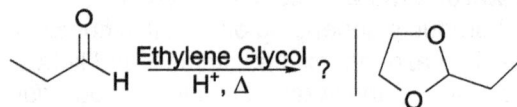

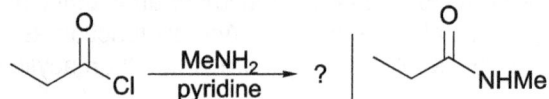

Figure 109: More reactions of aldehydes and ketones

Carboxylic acids, carboxylic acid derivatives, and nitriles

Carboxylic Acids, Carboxylic Acid Derivatives, and Nitriles as starting materials. Carboxylic acids and their derivatives are very common starting materials, as they can be interconverted among each other fairly easily. We will have a brief discussion of how these work and then address the reactions. The acid halide is the reactant that can easily be converted into all of the other acid derivatives via a nucleophilic acyl substitution reaction.

There are many derivatives, but the ones we will see here (listed by reactivity) are acid chlorides, acid anhydrides, esters, and amides. A derivative listed earlier can be converted to any derivative to the right of it. For example, esters can be easily converted into amides, but not easily converted into acid anhydrides or acid chlorides. First, we will see reactions that convert acid chlorides into acid anhydrides, esters, and amides in Figure 110.

(a) What is the product of this reaction?

(b) What is the product of this reaction?

(c) What is the product of this reaction?

Figure 110: Reactions of acid chlorides

In (a), we have propanoyl chloride reacting with sodium acetate. Acid chlorides generally react by substitution, so the acetate simply replaces the -Cl, and we get the product shown, acetic propanoic anhydride. In (b), we have propanoyl chloride reacting with methanol under basic conditions to give the ester, methyl propanoate. In (c), we have propanoyl chloride reacting with methyl amine under basic conditions to give N-methylpropanamide. Next, we will look at converting anhydrides into esters and amides and converting esters into amides in Figure 111.

(a) What is the product of this reaction?

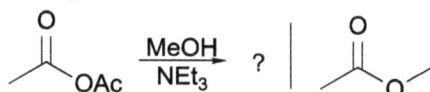

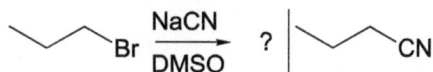

(b) What is the product of this reaction?

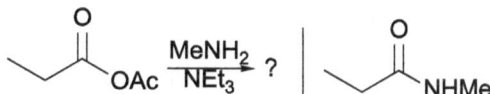

(c) What is the product of this reaction?

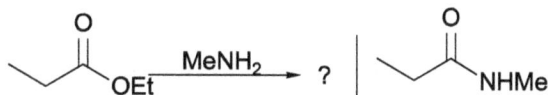

Figure 111: Reactions of acid anhydrides and esters

In (a), we have acetic anhydride reacting with methanol under basic conditions. This will proceed via an acyl substitution to give the ester, methyl acetate. In (b), we see that reacting the acetic anhydride with methylamine gives N-methyl propanamide. In (c), we are simply reacting methylamine with ethyl propanoate to also give the product N-methyl propanamide.

Nitriles can be made in three main ways: substitution reactions of alkyl halides, dehydration reactions of primary amides, and reactions of aldehydes/ketones with NaCN and acid. Examples of these are shown in Figure 112.

(a) What is the product of this reaction?

(b) What conditions are best for this reaction?

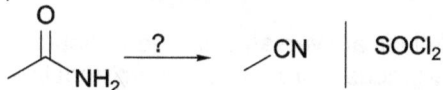

(c) What conditions are best for this reaction?

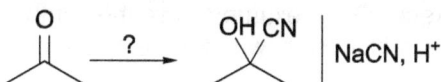

Figure 112: Reactions affording nitrile-containing compounds

In (a), we see the reaction of propyl bromide with sodium cyanide in DMSO. These are classic S_N2 reaction conditions, so in this case the product is butanenitrile. In (b), we see that we are trying to convert acetamide into acetonitrile. Here, the important thing to note is that only primary amides can be converted to nitriles with thionyl chloride. In (c), we see that we are converting a ketone into a cyanohydrin. Here, the best conditions are a mixture of acid and NaCN.

Conversion of nitriles to other functional groups. As mentioned, there are many interconversions between the carboxylic acid derivatives and nitriles. In addition, there are a number of reactions that can take nitriles and convert them into a variety of other functional groups. Some examples are shown in Figure 113.

(a) What is the product of this reaction?

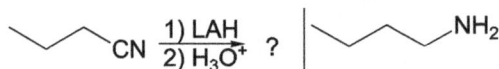

$$\text{CN} \xrightarrow[\text{2) H}_3\text{O}^+]{\text{1) LAH}} \text{?} \quad | \quad \text{NH}_2$$

(b) What conditions are best for this reaction?

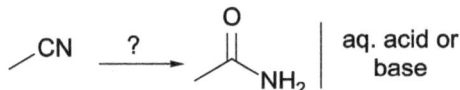

$$\text{CN} \xrightarrow{\text{?}} \underset{\text{NH}_2}{\overset{\text{O}}{\|}} \quad \bigg| \quad \begin{array}{c}\text{aq. acid or}\\ \text{base}\end{array}$$

(c) What conditions are best for this reaction?

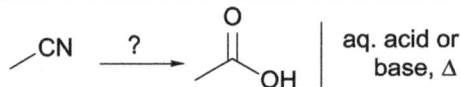

$$\text{CN} \xrightarrow{\text{?}} \underset{\text{OH}}{\overset{\text{O}}{\|}} \quad \bigg| \quad \begin{array}{c}\text{aq. acid or}\\ \text{base, }\Delta\end{array}$$

(d) What is the product of this reaction?

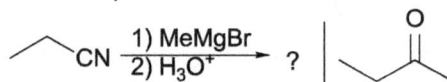

$$\text{CN} \xrightarrow[\text{2) H}_3\text{O}^+]{\text{1) MeMgBr}} \text{?} \quad \bigg| \quad \overset{\text{O}}{\|}$$

Figure 113: Conversion of nitriles into other functional groups

In (a), we are looking at the product of an LAH reduction of butanenitrile. The reduction of a nitrile will give the primary amine under these conditions, which is what we see in the product pentylamine. In (b) and (c), there is a slight difference in conditions-namely the presence, or lack thereof, of heat. The hydration of acetonitrile in (b) is quite easy, resulting in the stable amide. In (c), more energy is needed to convert the nitrile to the carboxylic acid since the amide is the most stable of the acid derivatives discussed here. In (d), we predict what happens when a nitrile reacts with a strong, basic nucleophilic reagent such as MeMgBr. The carbon of the nitrile is electrophilic, and the product after workup is a ketone.

Amines

Amines as starting materials. Amines are versatile nitrogen-containing compounds, but there are relatively few reactions we see in undergraduate organic chemistry that involve amines as starting materials. Some examples we have already seen in previous reactions, such as in Figure 92(c), which show an amine being used to make an alkene. Other examples are the synthesis of imines and enamines using an amine and an aldehyde/ketone, which we saw in Figure 108 and Figure 109.

Amines as products. Reactions that give amines as products are much more prevalent since they can be derived in at least three fashions: (1) Reductions/reductive aminations of imines or carbonyl compounds, (2) LAH reductions of amides, and (3) S_N2 reactions of alkyl halides with nitrogen nucleophiles followed by reduction or deprotection. We will see examples of these. Our first examples are shown in Figure 114.

(a) What starting material would be best for this reaction?

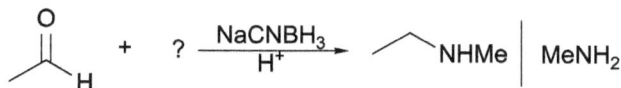

(b) What is the product of this reaction?

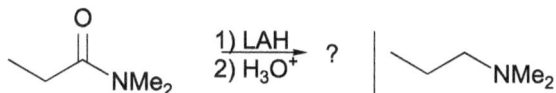

(c) What is the product of this reaction?

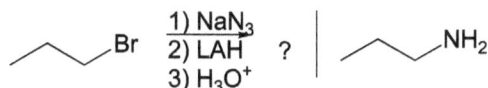

(d) What is the product of this reaction?

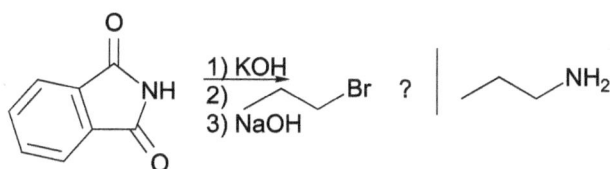

Figure 114: Reactions that give amines as products

In (a) we see are looking at the reductive amination (given away by the use of sodium cyanoborohydride) of acetaldehyde to give a secondary amine product. Here, the missing reactant is methylamine. In (b), we predict the product of the LAH reduction of a tertiary amide. Reductions of amides with LAH result in the carbonyl being reduced to the CH_2 after the reaction. Thus, the product is the tertiary amine shown. Examples (c) and (d) both show routes to synthesize primary amines, which cannot be easily synthesized and isolated because of the tendency of simple S_N2 reactions of alkyl halides with ammonia to give secondary amine products in significant amounts due to multiple alkylations. In (c), we have an S_N2 reaction with sodium azide followed by an LAH reduction to give the primary amine. In (d), we have what is called the Gabriel synthesis in that phthalimide is reacted first with KOH, followed by propyl bromide and finally KOH to generate the primary amine.

Two other ways to make amines via rearrangements are shown below in Figure 115.

(a) What is the product of this reaction?

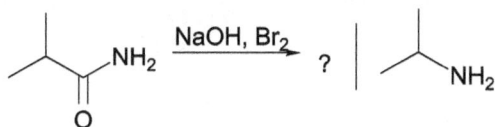

(b) What is the product of this reaction?

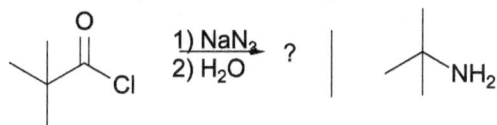

Figure 115: Rearrangements of acid derivatives to amines

In both (a) and (b) the overall reaction is to remove C=O from the molecule to give an amine that has one less carbon. In (a), in a reaction known as the Hofmann rearrangement, a primary amide reacts with sodium hydroxide and bromine. The result is the propan-2-amine shown. In (b), we see a related reaction known as the Curtius rearrangment in which an acid chloride is reacted with sodium azide, followed by water to give 2-methylpropane-2-amine as the product. One good feature of these reactions is that sterically hindered amines can be generated.

Synthesis

Synthesis is a sequence of two or more reactions. In organic chemistry, there are a very large number of possible syntheses. For example, common topics found in undergraduate organic chemistry courses are the acetoacetic ester synthesis and malonic ester synthesis, named for the starting materials in those sequences. Most reaction sequences don't have particular names since there are so many variations. This chapter will be devoted to two- or three-step sequences, since all the other sequences can be made up of multiple two- or three step-sequences. Again, there are many different formats for these questions; three are shown below.

1. **Multiple choice.** Here, a sequence of two or more reactions give a particular product.

2. **A roadmap or fill in.** A short answer-type of question where some conditions and products in a sequence are given, and others need to be filled in. Here, it is important to use the known/given information to help.

3. **Open response.** These are open-ended questions regarding synthesis (i.e., starting with styrene, synthesize benzoic acid).

It should also be noted that these questions may or may not involve practical reaction sequences. We will first start with typical reaction sequences where each functional group is a starting material and a product. We will notice that some functional groups (e.g., alcohols, ketones, aldehydes) will have more reactions discussed than others (e.g., alkanes). There are a very large number of possible reactions; this chapter will focus on some of the more commmon sequences seen with reactions from organic chemistry undergraduate courses. This is not meant to be a comprehensive list of reactions but should cover many of them.

Alkanes

Alkanes are present in most organic molecules, yet there are not a large number of sequences that produce alkanes as products. Most reaction sequences that produce alkanes involve a reduction reaction of some kind. We can see three examples of sequences in Figure 116.

In (a) we have the common reaction of an alkylation followed by reduction. In particular, we have the alkylation of butyne followed by reduction with H_2 and Pd/C to give pentane. In (b), we have the Wittig reaction of 2-methylpropanal followed by reduction of the resulting alkene with H_2 and Pd/C to give 2-methylbutane. In (c), we have radical addition of HBr to 2-methylbut-1-ene followed by reaction with a dimethylcuprate to give 3-methylpentane as the product.

(a) 1) NaH 2) MeBr 3) Pd/C, H₂

(b) 1) =PPh₃ 2) H₂, Pd/C

(c) 1) HBr, ROOR 2) (Me)₂CuLi

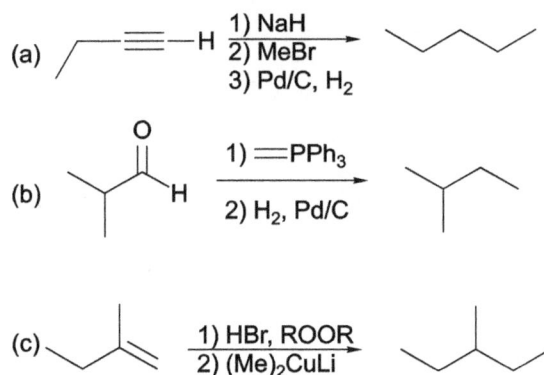

Figure 116: Example syntheses of alkanes

Alkenes

Alkenes can be synthesized via a number of methods: (1) reaction of aldehydes or ketones with phosphorus ylides, (2) dehydration of alcohols, (3) elimination of alkyl halides, and (4) from reduction of alkynes. Examples of the first three are shown below in Figure 117.

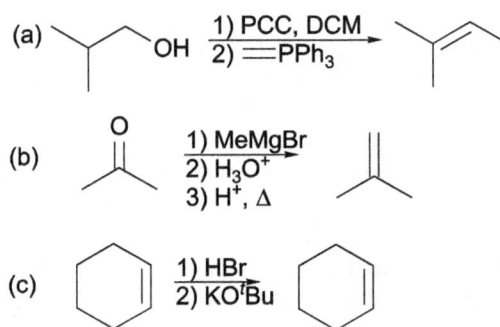

(a) OH 1) PCC, DCM 2) =PPh₃

(b) 1) MeMgBr 2) H₃O⁺ 3) H⁺, Δ

(c) 1) HBr 2) KOᵗBu

Figure 117: Syntheses of alkenes

In (a), we see 2-methylpropan-1-ol being oxidized to the aldehyde by PCC in DCM followed by a Wittig reaction to give 2-methylbut-2-ene as the product. In (b), we see acetone being first reacted with a methyl Grignard followed by dilute acid workup and strong acid catalysis to dehydrate and give 2-methylprop-1-ene as the product. In the third example, we see the addition of HBr to cyclohexene to give bromocyclohexane followed by E2 elimination with a strong base to again give cyclohexene as the product. Two more routes to make alkenes are shown in Figure 118.

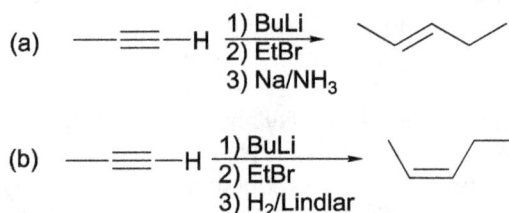

(a) —≡—H 1) BuLi 2) EtBr 3) Na/NH₃

(b) —≡—H 1) BuLi 2) EtBr 3) H₂/Lindlar

Figure 118: Syntheses of alkenes by partial reduction of alkynes

These examples showcase the different reduction methods of alkynes. In (a), we have the reaction of prop-1-yne with butyl lithium followed by alkylation to give pent-2-yne. In the third step in (a) we have the dissolving metal reduction with sodium in ammonia to give the final product, (2E)-pent-2-ene. In (b), we see a similar sequence, except the final step is H_2 and a modified catalyst to give (2Z)-pent-2-ene.

Alkynes

Alkynes can be synthesized by two main methods: (1) alkylation of acetylene or terminal alkynes and (2) elimination reactions. Examples of both are shown in Figure 119.

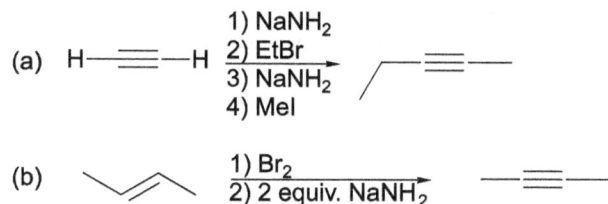

(a)   H———≡———H   1) NaNH$_2$ 2) EtBr 3) NaNH$_2$ 4) MeI

(b)   1) Br$_2$ 2) 2 equiv. NaNH$_2$

Figure 119: Syntheses of alkynes

In (a), we have acetylene being deprotonated and alkylated twice, first to add an ethyl group, and second to add a methyl group, resulting in the product pent-2-yne. It should be noted that the pKa of the Csp-H bonds in an alkyne are more acidic than other hydrocarbons. In (b), but-2-yne is generated by reacting (2E)-but-2-yne with bromine followed by a double elimination with sodium amide.

Alkyl Halides

Alkyl halides can be generated by a few main methods: (1) conversion of alcohols, (2) reactions of alkenes and alkynes with acids, (3) radical halogenation of alkanes, and (4) allylic bromination. We can see examples of the first two in Figure 120:

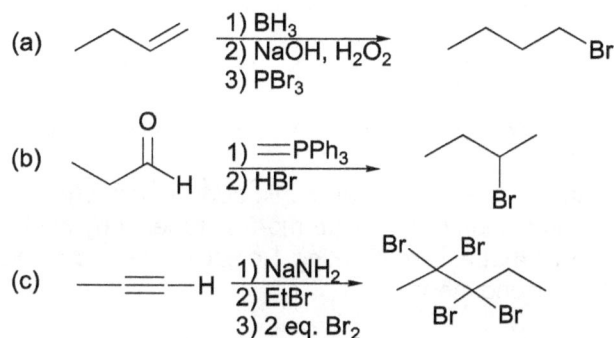

(a)   1) BH$_3$ 2) NaOH, H$_2$O$_2$ 3) PBr$_3$

(b)   1) =PPh$_3$ 2) HBr

(c)   ———≡———H   1) NaNH$_2$ 2) EtBr 3) 2 eq. Br$_2$

Figure 120: Syntheses of alkyl halides

In (a), we see that we are generating butan-1-ol by performing the hydroboration-oxidation sequence on but-1-ene. This is followed by the reaction with the primary alcohol with phosphorous tribromide to give 1-bromobutane as the product. In (b), we are first performing a Wittig reaction on propanal followed by the addition of HBr to give 2-bromobutane, following Markovnikov's rule. In (c), we see prop-1-yne being deprotonated and alkylated to give pent-2-yne, followed by tetrabromination with two equivalents of bromine to give 2,2,3,3-tetrabromopentane as the product.

We can see two more examples in Figure 121.

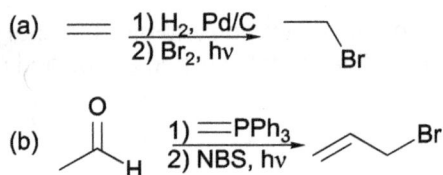

Figure 121: More syntheses of alkyl halides

In the first example, we see the reduction of ethylene to ethane followed by radical halogenation to give bromoethane as the product. In the second case, we see that we are reacting acetaldehyde via the Wittig reaction followed by allylic bromination with *N*-bromosuccinimide and light to give 3-bromoprop-1-ene, also known as allyl bromide. Note that in these two examples we are using radical chemistry to substitute a C-H bond.

Aromatics

Aromatic compounds are generally made from other aromatic compounds via substitution reactions. Thus, there are two types: (1) electrophilic aromatic substitution (EAS) and (2) nucleophilic aromatic substitution (NAS). We see examples of reactions below in Figure 122:

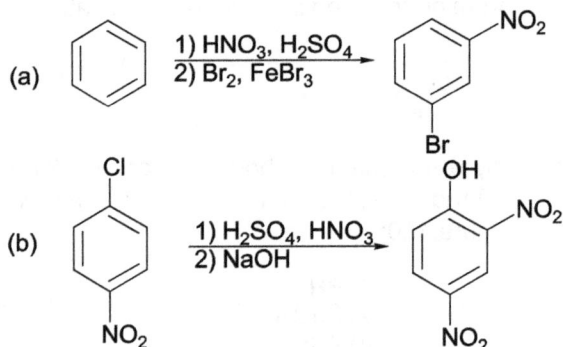

Figure 122: Syntheses involving aromatic compounds

In (a), we have an example of electrophilic aromatic substitution where the two step sequence involves nitration followed by bromination to give 1-bromo-3-nitrobenzene as the product. The nitro group is a meta-director. In (b), we see that we are nitrating 1-chloro-4-nitrobenzene followed by the substitution of the -Cl with -OH in the second step.

Alcohols and Phenols

As with alkenes, there are a number of ways to make alcohols: (1) Hydration of alkenes, (2) reduction of carbonyl compounds, (3) solvolysis of alkyl halides, (4) addition of Grignards/alkyl lithiums to carboxylic acid derivatives. Examples are shown in Figure 123.

In (a), we have a four-step sequence starting with oxidizing 2-methyl-propan-1-ol to the aldehyde, followed by a Wittig reaction. The resulting alkene is then subjected to the hydroboration-oxidation sequence to give the anti-Markovnikov product 3-methylbutan-1-ol. In (b), we have acetyl chloride being converted into an ester with methanol under basic conditions, followed by reduction with LAH and acid

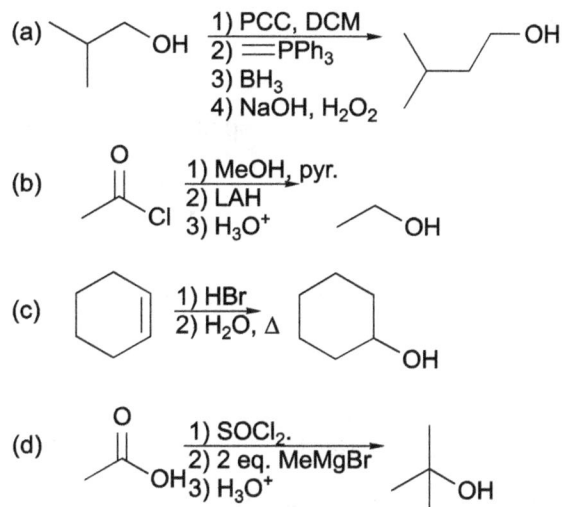

Figure 123: Syntheses of alcohols

workup to give ethanol as the product. In (c), we have addition of HBr to cyclohexene followed by hydrolysis to give cyclohexanol as the product. Finally, in (d), we have acetic acid being converted to its acid chloride in step 1, followed by reaction with two equivalents of methyl magnesium bromide followed by acid workup to give 2-methylbutan-2-ol as the product.

Ethers

Ethers can be formed in a few ways: (1) S_N2 attack of an alkoxide on an alkyl halide, (2) nucleophilic attack of an alcohol on a carbocation, and (3) use of an alcohol solvent when reacting alkenes with bromine. We will see examples below in Figure 124.

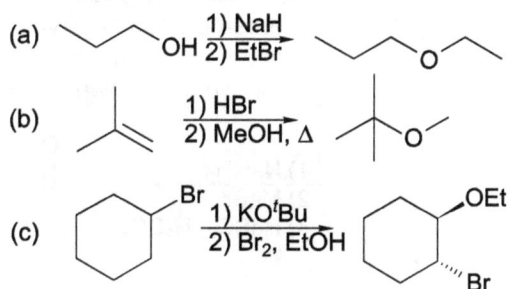

Figure 124: Syntheses of ethers

In (a), we have the classic ether synthesis involving the S_N2 reaction of the alkoxide derived from propan-1-ol with ethyl bromide to give the product ethyl propyl ether. In (b), we have the reaction of 2-methylpropene with HBr to give the alkyl bromide, followed by an S_N1 reaction with methanol to give 2-methoxy-2-methylpropane as the product. In (c), we perform an E2 reaction to give cyclohexene, formed by using EtOH instead of H_2O to give a bromo ether.

Thiols and Sulfides

Thiols can be synthesized a couple of different ways, both involving S_N2 reactions of alkyl halides with either sodium hydrosulide (this method suffers from competing sulfide formation), or thiourea followed by hydrolysis. Sulfides are mainly synthesized by the S_N2 reactions of thiols with alkyl halides. Examples are shown in Figure 125.

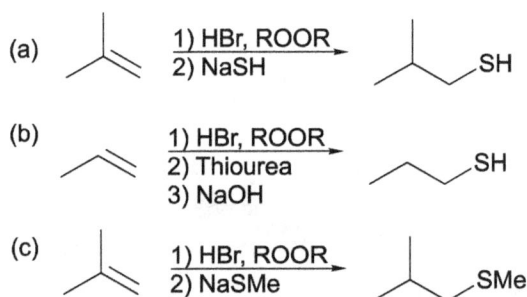

Figure 125: Syntheses of thiols and sulfides

In (a), we have the radical addition of HBr to 2-methylprop-1-ene followed by the S_N2 reaction of sodium hydrosulfide to give 2-methylpropan-1-thiol. In (b), we have a similar sequence, starting with prop-2-ene. After the radical addition of HBr, thiourea is used as a nucleophile to form the new C-S bond, and the sodium hydroxide is used to generate the product propan-1-thiol. In (c), we have the radical addition of HBr to 2-methylprop-1-ene followed by the S_N2 reaction of sodium methanethiolate to give 2-methyl-1-(methylsulfanyl)propane as the product.

Aldehydes and Ketones

There are a large number of reactions that can generate aldehydes and ketones. These are as follows: (1) hydroboration-oxidation and oxymercuration-demercuration of alkynes, (2) ozonolysis of alkenes, (3) oxidative cleavage of diols, (4) oxidation of alcohols, and (5) attack of alkyl lithiums/alkyl Grignards on nitriles, followed by hydrolysis. Examples of the first two are shown in Figure 126.

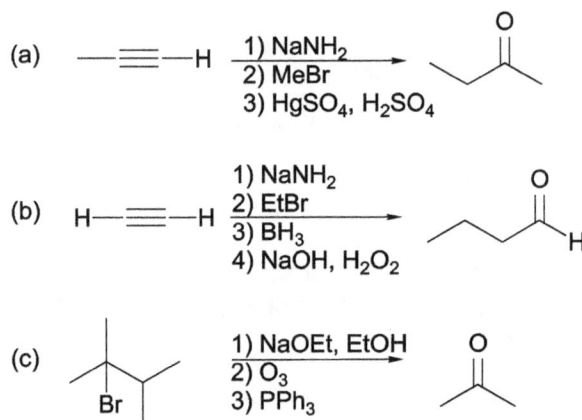

Figure 126: Syntheses of aldehydes and ketones

In (a), we have a sequence where we react acetylene with sodium amide followed by methyl bromide to give the four-carbon alkyne. This is followed by mercury-catalyzed hydration to give butan-2-one as the product. In (b), we are starting with acetylene, reacting it with sodium amide followed by alkylation. The resulting product is then subjected to hydroboration-oxidation to give the product butanal. In (c), we are starting with the alkyl halide 2-bromo-2,3-dimethylbutane. The first step generates a tetrasubstituted alkene via elimination. The alkene is then subjected to ozonlysis followed by quenching with triphenylphosphine to give acetone as the organic product of the reaction. Next, we will look at oxidations and reactions of nitriles in Figure 127.

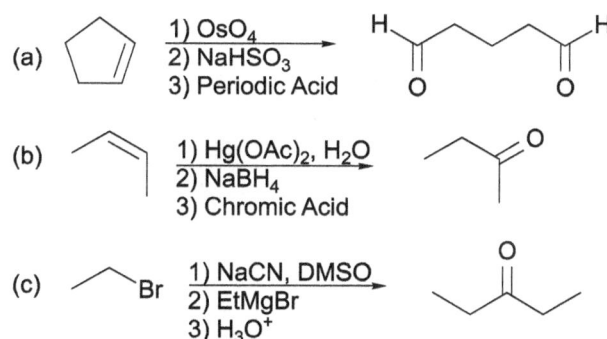

Figure 127: More syntheses of aldehydes and ketones

In (a), we are starting with cyclopentene. It is first reacted with osmium tetroxide followed by a workup with sodium bisulfite to give a diol intermediate, which is reacted in the last step with periodic acid to give the dialdehyde product shown. In (b), we are taking (2Z)-but-2-ene and performing the oxymercuration-demercuration sequence followed by oxidation with chromic acid to give butan-2-one as the product. Finally, in (c), we are reacting ethyl bromide with sodium cyanide in DMSO to first give propanenitrile. This is then reacted with ethyl magnesium bromide to give pentan-3-one after acid workup.

Carboxylic Acids, Carboxylic Acid Derivatives, and Nitriles

As we saw previously, there are a large number of interconversions among the carboxylic acid derivatives, as well as a few separate reactions that can directly give these functional groups. Thus, the main classes of reactions that will give these products are (1) Conversion of acid chlorides to other acid derivatives, (2) conversion of other acid derivatives among each other, (3) conversion of alkyl halides to acid derivatives via substitution reactions, and (4) addition of Grignard reagents to carbon dioxide. Let's look at a few examples in Figure 128.

In (a), we start with benzoic acid, which is first converted to its acid chloride with thionyl chloride. The acid chloride is then reacted with two equivalents of methylamine to give the product N-methylbenzamide. In (b), we have a similar sequence, but here we are starting with the acetyl chloride. This is converted to the ester with sodium methoxide and then reacted with dimethylamine and heat to give the product N,N-dimethylacetamide as the product. In (c), radical addition of HBr to prop-1-ene is followed by S_N2 reaction with sodium cyanide in DMSO to give propanenitrile as the product. Finally, in (d) we have the addition of HBr to prop-2-ene followed by conversion into a Grignard reagent with elemental magnesium. The Grignard is then reacted with CO_2 to give 2-methylpropanoic acid after workup.

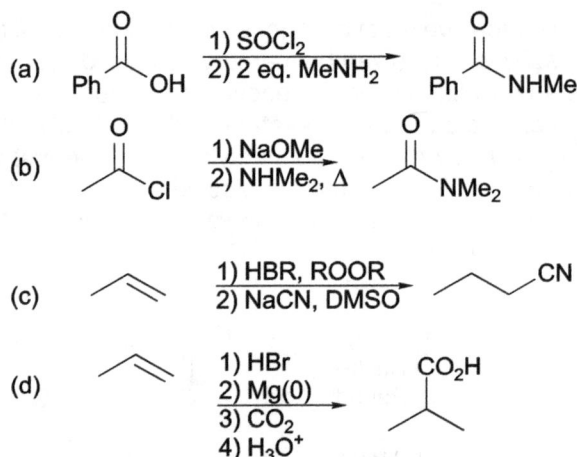

Figure 128: Syntheses producing carboxylic acids and derivatives

Amines

Amines can be synthesized a number of ways, including by reduction of amides, substitution reactions with proper nucleophiles, and reduction of imines/reductive amination. When we consider two-step syntheses, a number of possibilities exist, and we will go through a few common ones in Figure 129.

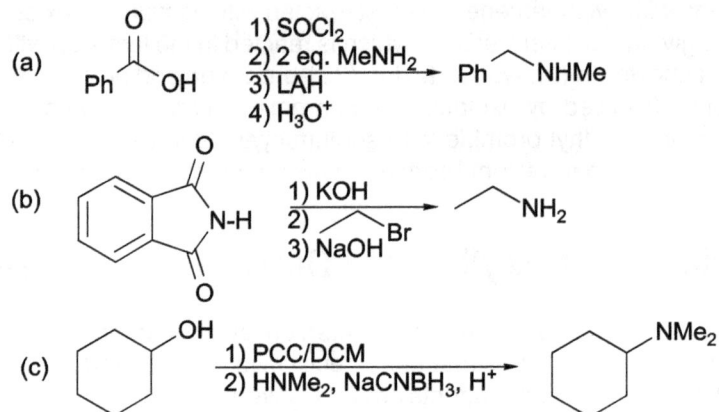

Figure 129: Syntheses producing amines

In (a), we start with benzoic acid, which we are converting to its acid chloride. This is then reacted with two equivalents of methylamine to give the amide derivative, which is then reduced to the secondary amine using LAH. In (b), we have an example of the Gabriel synthesis using ethyl bromide to give ethylamine as the product. Finally, in (c) we have an example of a reductive amination. Cyclohexanol is first oxidized to cyclohexanone using PCC. The resulting ketone then undergoes reductive amination to give the product, *N,N*-dimethylcyclohexanamine.

Worked Example Questions

For the final section of the book, where will be 10 synthesis questions from the types listed below. Note that there are a vast variety of questions that involve synthesis that are not in these categories, but these are some common types.

1. **Roadmap.** These types of questions have a partial multi-step synthesis and ask for product structures or conditions.

2. **Multiple choice.** These are typical multiple-choice questions, but there are multiple reaction steps to sort out.

3. **Predict the product** (open response). These are multiple step questions, but you generate an answer instead of choosing one.

4. **Synthesis proposal.** These types of questions indicate a target molecule that needs to be synthesized. Typically, there are some conditions attached (number of steps, starting materials, types of reactions that can be used). After each type of question, some hints will be given. Finally, answers will be given.

Questions 1-5 are shown in Figure 130.

For question 1, we need to consider what happens when HBr is reacted with an alkene. In the absence of peroxides, the bromine will add to the more substituted carbon of the alkene in accordance with Markovnikov's rule. Here, we have peroxides, which switches the product to where the bromine ends up on the less substituted carbon of the alkene. Structure **A** is going to depend on the regiochemistry of the addition. For **B** we need to look at what is changing. For structure **C**, we need to consider what happens when a tertiary alcohol is heated in sulfuric acid.

For question 2, we need to look at the first set of conditions and what that adds to compounds. For conditions **B** and structure **C**, we need to look at the final structure and consider what is happening.

For question 3, for conditions **A** and structure **B**, we need to look at the amide product and consider which reactions convert an acid to an amide. For conditions **C**, we need to recognize which conditions can reduce an amide to an amine.

For the multiple-choice questions 4 and 5, we need to consider what is going on in each step and think about the functional groups that are the products of each of those steps.

Five other questions are presented in Figure 131.

Questions 6 – 8 are similar to questions 4 and 5 except you are required to generate the answer instead of choose it. The process of answering is similar, considering what is going on in each step. Ideally, you would draw out the product after each step to help keep things straight in your mind.

Questions 9 and 10 are among the most challenging because you must consider a large number of potential reaction sequences and all of the issues/concerns that come with them.

Finally, we will look at the answers and do a full analysis of the questions. The answer to questions 1 - 5 are shown in Figure 132.

For question 1, the first reaction is the addition of HBr in the presence of peroxides to an alkene. In this case, the HBr adds with anti-Markovnikov orientation to give propyl bromide **A**. Looking at the next known structure (a teriary alcohol), we can work backwards and imagine that we had our propyl group added to acetone. Thus, the three steps of **B** are (1) Mg(0) (to make the Grignard), (2) and (3) addition of that Grignard to acetone to give the tertiary product alcohol after workup. Finally, the heated sulfuric acid conditions will dehydrate the tertiary alcohol to give the trisubstituted alkene product **C**.

For question 2, the first boxed structure **A** is diethyl malonate. This question is an example of the malonic ester synthesis, named for the starting material. The first set of conditions will lead to deprotonation (NaOMe), followed by alkylation (MeI). The second set of conditions **B** is very similar to the first, although we can tell from the final carboxylic acid product that an ethyl group has been added. Thus, the other alkyl group added is an ethyl group. The final step of the reaction (acid and heat) will decarboxylate diester **C**.

For question 3, the reaction starts with a carboxylic acid and then progresses to a tertiary amide after reacting with a secondary amine. The type of reaction that will produce the amide is a reaction of an amine with an acid chloride. Thus, the first reactant **A** is thionyl chloride, and the first product **B** is acetyl chloride. The final step is a reduction of an amide to an amine, which is accomplished with LAH followed by an acid workup **C**.

For question 4, the starting material is a terminal alkene. The first two steps are the hydroboration/oxidation reaction, which is going to give the primary alcohol as a product due to the mechanism of the reaction. Thus, we can say our product will either be choice B or choice C. The third step, chromic acid, is known to oxidize primary alcohols to the carboxylic acid. Thus, our answer is **B**.

For question 5, we start with an ester, and the first step is an acidic hydrolysis to give the carboxylic acid. Step 2 converts the acid to the acid chloride. Steps 3 and 4 are part of the Curtius rearrangement, which converts an acid (or acid chloride) to an amine with one less carbon, since carbon dioxide is given off as a byproduct. Since our initial carboxylic acid had three carbons, our product will have two carbons. Thus, the answer is **A**.

Finally, we have the answers and analysis for questions 6 – 10 below in Figure 133.

For question 6, we have a portion of question 1 asked a bit differently. Here, we start with propyl bromide. After step 1, we have the propyl Grignard reagent. After steps 2 and 3, we have a tertiary alcohol product, and step 4 gives the trisubstituted alkene as the product.

For question 7, we start with cyclohexene. The first step will generate an epoxide, and the final two steps involve a ring-opening nucleophilic attack via an S_N2 process followed by a workup to give the trans-2-methylcyclohexanol as the product.

For question 8, we start with a terminal alkyne. Terminal alkynes have a significantly acidic Csp-H bond, which can be deprotonated with a strong base like $NaNH_2$. After deprotonation, we add ethyl bromide to give an internal alkyne. This alkyne is then reduced with H_2/Lindlar to give the alkene product shown.

For question 9, we are tasked with starting from prop-1-yne and synthesizing (2E)-hex-2-ene. Looking at our product, we know we have done the following: (1) added three carbons, and (2) converted an alkyne to a trans-alkene (also an E)-alkene in this case). Thus, we need to think about which we want to do first, make the C-C bond, or reduce the alkyne to the alkene. As we saw in question 8, the Csp-H bond is quite acidic, so the better route is to deprotonate and make the new C-C bond first. If we reduce first, we have an issue of trying to selectively deprotonate an alkene. This is quite a bit more difficult for two reasons: (1) the Csp^2-H is significantly less acidic than the Csp-H, and (2) there are three Csp^2-H bonds to contend with instead of one Csp-H in the alkyne. Thus, we want to make the new C-C bond first, and a sequence of deprotonation with sodium amide followed by alkylation with propyl bromide will give the internal alkyne. This can then be reduced with sodium in ammonia to give the trans or *E* alkene.

Finally, for question 10, we are tasked with providing two syntheses of triethylamine using an amine in the process. For making tertiary amines, two sequences that involve amines are shown. In A, a reductive amination is proposed. Since a reductive amination makes a new C-N bond, our carbonyl component needs to have two carbons, so acetaldehyde is used as our carbonyl component. Diethylamine is used as our other starting material. When these are mixed together in the presence of sodium cyanoborohydride, triethylamine is generated by reduction. The cyanoborohydride is less reactive and selectively attacks the iminium intermediate in preference to the aldehyde. The second route involves generating a C-N bond by making an amide, followed by reduction.

1) Fill in the indicated conditions or structures for the following synthesis:

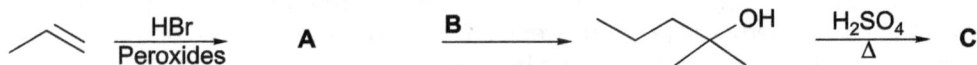

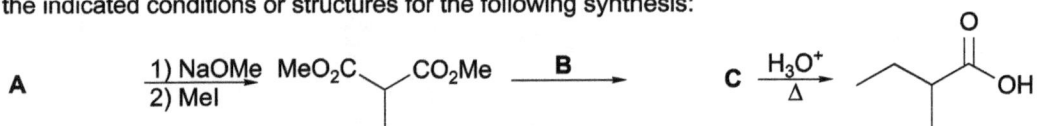

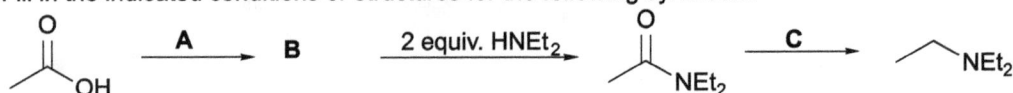

2) Fill in the indicated conditions or structures for the following synthesis:

A   1) NaOMe   MeO$_2$C   CO$_2$Me   **B**   **C** $\xrightarrow{H_3O^+}$ OH
     2) MeI

3) Fill in the indicated conditions or structures for the following synthesis:

4) What is the product of this reaction sequence?

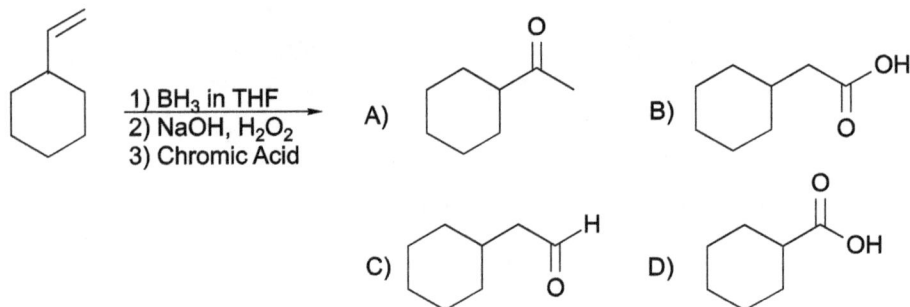

1) BH$_3$ in THF
2) NaOH, H$_2$O$_2$
3) Chromic Acid

A)   B)   C)   D)

5) What is the product of this reaction sequence?

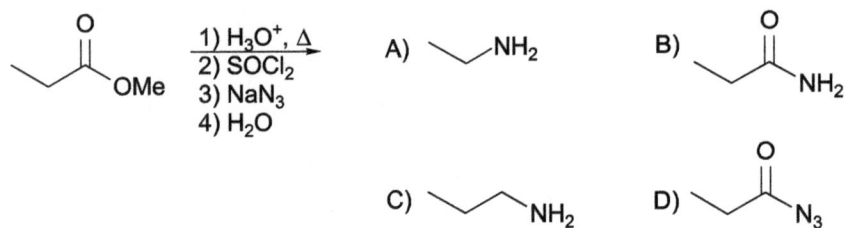

1) H$_3$O$^+$, Δ
2) SOCl$_2$
3) NaN$_3$
4) H$_2$O

A) $\diagup$NH$_2$ B)   NH$_2$

C)   NH$_2$ D)   N$_3$

Figure 130: Synthesis questions 1 - 5

6) What is the product of this reaction sequence?

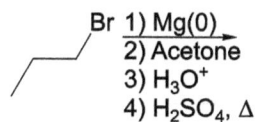

Br 1) Mg(0)
    2) Acetone
    3) H_3O^+
    4) H_2SO_4, Δ

7) What is the product of this reaction sequence?

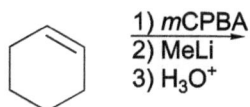

1) *m*CPBA
2) MeLi
3) H_3O^+

8) What is the product of this reaction sequence?

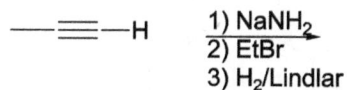

—≡—H   1) $NaNH_2$
        2) EtBr
        3) H_2/Lindlar

9) Starting from prop-1-yne, synthesize (2*E*)-hex-2-ene. You may use reagents as needed.

10) Synthesize triethylamine two different ways.
    You must use an amine but can use other reagents as needed.

Figure 131: Synthesis questions 6 - 10

1) Fill in the indicated conditions or structures for the following synthesis:

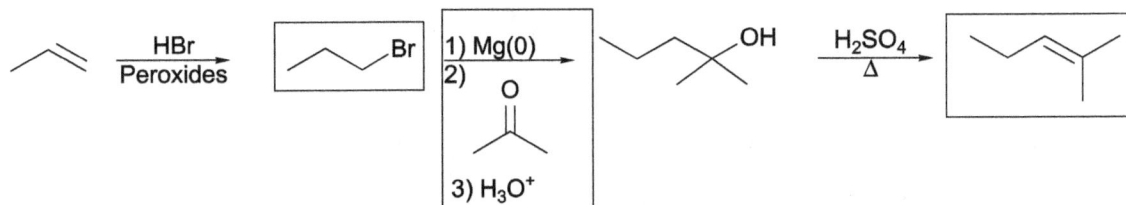

2) Fill in the indicated conditions or structures for the following synthesis:

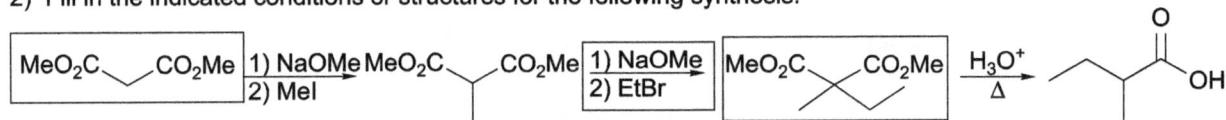

3) Fill in the indicated conditions or structures for the following synthesis:

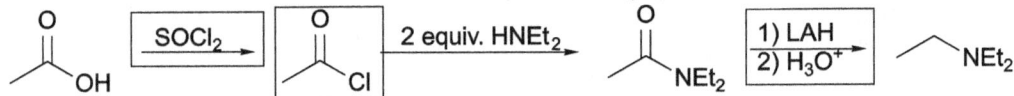

4) What is the product of this reaction sequence?

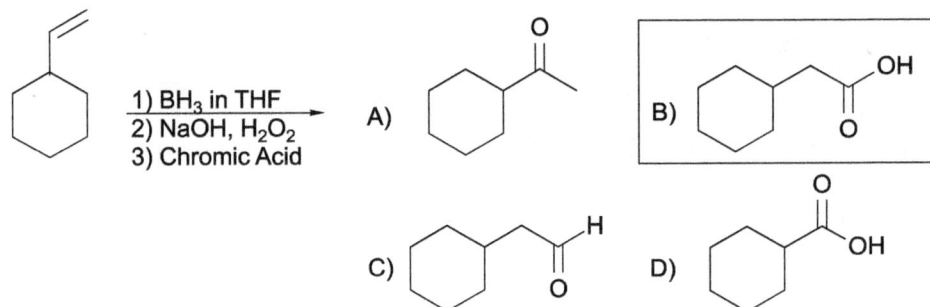

1) BH_3 in THF
2) NaOH, H_2O_2
3) Chromic Acid

A)   B)   C)   D)

5) What is the product of this reaction sequence?

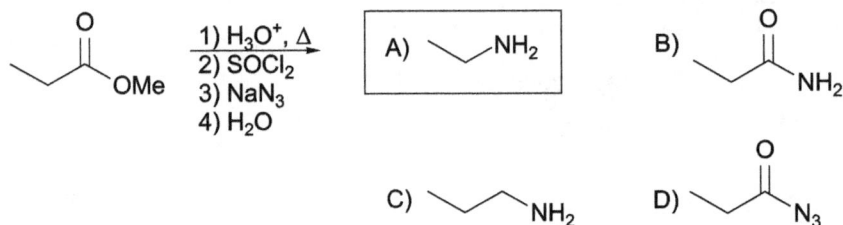

1) H_3O^+, Δ
2) $SOCl_2$
3) NaN_3
4) H_2O

A)   B)   C)   D)

Figure 132: Answers to synthesis questions 1-5

6) What is the product of this reaction sequence?

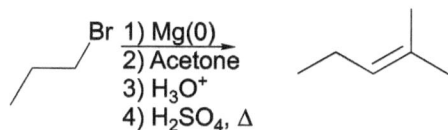

Br 1) Mg(0)
2) Acetone
3) H_3O^+
4) H_2SO_4, Δ

7) What is the product of this reaction sequence?

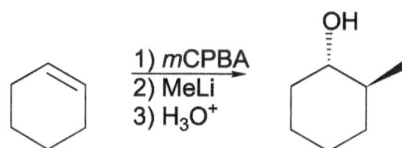

OH

1) mCPBA
2) MeLi
3) H_3O^+

8) What is the product of this reaction sequence?

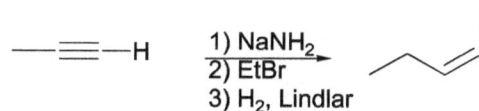

—≡—H

1) $NaNH_2$
2) EtBr
3) H_2, Lindlar

9) Starting from prop-1-yne, synthesize (2E)-hex-2-ene. You may use reagents as needed.

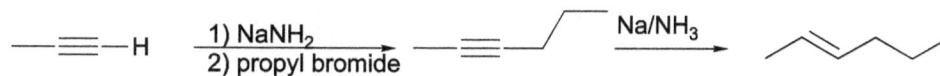

—≡—H

1) $NaNH_2$
2) propyl bromide

—≡—

Na/NH_3

10) Synthesize triethylamine two different ways.
    You must use an amine, but can use other reagents as needed.

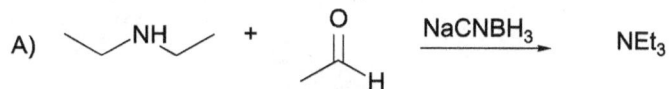

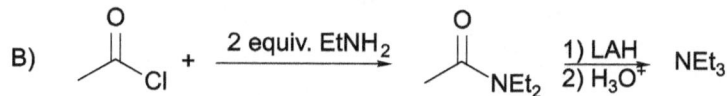

A) NH +

O

H

NaCNBH$_3$

NEt_3

B)

O

Cl

+

2 equiv. EtNH$_2$

O

NEt$_2$

1) LAH
2) H_3O^+

NEt_3

Figure 133: Answers to synthesis questions 6-10

Credit